AFFIRMATIVE DISCRIMINATION: ETHNIC INEQUALITY AND PUBLIC POLICY

BOOKS BY NATHAN GLAZER

(with David Riesman and Reuel Denney) *The Lonely Crowd*
(with David Riesman) *Faces in the Crowd*
American Judaism
The Social Basis of American Communism
Beyond the Melting Pot (with Daniel P. Moynihan)
Remembering the Answers
Affirmative Discrimination

BOOKS EDITED BY NATHAN GLAZER

Studies in Housing and Minority Groups (with Davis McEntire)
Ethnicity: Theory and Experience (with Daniel P. Moynihan)
The Urban Predicament (with William Gorham)

AFFIRMATIVE

DISCRIMINATION:

ETHNIC

INEQUALITY

AND

PUBLIC POLICY

Nathan Glazer

Basic Books, Inc., Publishers   New York

CONTENTS

Introduction: 1978   ix

Acknowledgments   xix

CHAPTER 1

The Emergence of an American Ethnic Pattern   3

CHAPTER 2

Affirmative Action in Employment: From Equal
    Opportunity to Statistical Parity   33

CHAPTER 3

Affirmative Action in Education: The Issue of
    Busing   77

CHAPTER 4

Affirmative Action in Housing: Overcoming
    Residential Segregation   130

CHAPTER 5

The White Ethnic Political Reaction   168

CHAPTER 6

Morality, Politics, and the Future of
    Affirmative Action   196

Notes   223

Index   235

INTRODUCTION: 1978

In *Affirmative Discrimination* I attacked the justice and the wisdom of shifting from individual rights to group rights in devising policies to overcome racial and ethnic-group discrimination and its heritage. Since the book was published, the issue has grown to one of giant proportions. As I write, a local school board in New York City has been superseded by the city's Board of Education because it refuses to provide ethnic and racial statistics to the Department of Health, Education, and Welfare; a judge in Detroit has declared quotas for the appointment of policemen unconstitutional; a contractor is suing the Massachusetts Bay Transportation Authority for rejecting his low bid in favor of one which would "set-aside" a higher proportion of the contract for minority subcontractors; the federal government is expanding the reach of set aside requirements in government contracts for minority-owned businesses; Los Angeles is preparing the largest school busing program yet attempted, following a state judge's order; Cleveland is preparing for massive school busing under a federal judge's orders; and the Supreme Court is pondering the case of Allan Bakke, a rejected applicant to the University of California at Davis Medical School who has been upheld by the Supreme Court of California in his claim that he was subjected to unconstitutional discrimination because of the medical school's quota for minorities.

The distribution of jobs and promotions on the basis of race and ethnic group, as described in chapter 2, has spread, by means of administrative agency requirements, to cover an ever larger part of the work force; school busing on the basis of race, as described in chapter 3, has spread to more communities. But recent decisions of the Supreme Court have given promise that those apparently inexorable developments may be slowed. Thus, tests that one would have expected, on the basis of previous court decisions and administrative practice, to be held discriminatory because average scores of persons of different races differed, have been upheld (*Washington* v. *Davis*, 1976, and other decisions). The full sweep of these decisions is uncertain; the administrative agencies persist in applying rules that would make most of these tests "discriminatory." The Supreme Court, which had apparently fashioned a rule with its *Keyes* decision in Denver, from which no school district could escape a finding of having practiced de jure segregation and thereby be made subject to massive busing, has again, in a somewhat uncertain and ragged course, withdrawn from its most forward position. In *Milliken* it refused to allow a metropolitan-wide remedy in Detroit; but it allows metropolitan-wide remedies in Wilmington and Indianapolis to go forward. In Dayton, it has said that remedies—and, of course, the issue is busing—must be in scale with the discriminatory acts demonstrated; but it has allowed busing in other communities with records that seem little different from Dayton to go forward without review. Affirmative action in housing and development, discussed in chapter 4, has had a rocky course in the federal courts. The Supreme Court has resisted the general claim that communities must try to achieve some appropriate racial and ethnic mix through positive action. State courts have been more receptive to this argument. But the Department of Housing and Urban Development may be expected to continue to push for such affirmative action through its regulations.

No brief sketch can hope to update the status of affirmative discrimination in all fields; every day there are hundreds of rulings by government agents, any of which may affect thousands or tens of thousands of employees, job applicants, and job holders, thousands or tens of thousands of students and teachers, hundreds of millions of dollars of contracts awarded on the basis of race, color, national origin, and ethnicity. A person's color, race, and ethnic background are as decisive for his or her fate today as they were in 1964, when, with the passage of the Civil Rights Act, we thought we had taken a giant step in eliminating color, race, and ethnic origin from public and private decision making. Matching the hundreds of millions being spent each year by government to impose goals and quotas and to monitor the actions of private and public employers, schools, and colleges, are the hundreds of millions spent by employers and institutions in their effort to comply, and now involving thousands of lawyers who have come to specialize in this arcane world of race and ethnic group requirement and privilege. That it could all be swept away, that we could achieve a society in which men and women are judged on the basis of their abilities rather than their color, race, or ethnic origin, now seems a utopia that is hopeless to expect or try to achieve. All one can hope for, or do, is to moderate the steady march toward a society in which race and ethnicity, stamped on everyone's identity card, are simply taken as a matter-of-course basis for judgement. (In some communities, individuals trying to escape being sent to a distant school by means of claiming one racial or ethnic identity are called up before expert boards for visual inspection and cross-examination to determine if they are "really" of the group they claim to be.)

It is scarcely conceivable that the Supreme Court, whatever it decides in the Bakke and other cases, can sweep all this away —the thousands of jobs, the hundreds of millions in budgets, the thousands of pages of administrative rulings, the millions of individual expectations, all of which add up to a society that habitually rules on people's fates on the basis of their race and ethnic group. But why can the Court not do so, when the Constitution, at least as interpreted the way it was until the early 1970s, and federal civil-rights legislation seem on the face of it to call for no less?

There are two reasons. The first is the continuing *condition* of the black population of the United States. I speak of condition, not of rights, nor even of the practice which affects them. Admittedly there are other racial and minority groups in the United States, but it is not their fate which motivates this massive machinery. Japanese and Chinese have moved ahead despite discrimination. Filipinos, Koreans, Asian Indians, Vietnamese, and other Asian immigrants can probably be expected to do the same, at different rates and with different problems. In any case, it is not their condition or political weight that maintains existing race conscious policies. The "Hispanic Americans" are a very mixed bag indeed, from the upwardly mobile Cubans to the depressed Puerto Ricans, but I doubt whether we would have employed race- and group-conscious policies on the scale we have employed them to deal with the problems of "Hispanic Americans." It is the blacks that are on the conscience of America, and quite properly: They were enslaved and rigidly kept down after freedom by massive public and private discrimination and prejudice. If, despite their grim history, they had made rapid progress, we would undoubtedly have not found it necessary to institute race-conscious policies in employment and education.

One should not underestimate the degree of black progress. From some perspectives it has been enormous: The number of blacks in professional jobs, in white collar jobs, earning high incomes, attending colleges and professional schools of some types, has increased rapidly. But the continued existence of a huge population of blacks of poor education, with no jobs or poor jobs, living in central cities from which jobs, middle class people, and services have fled—or been driven away—and in which crime and arson have become uncontrollable plagues, creates a demand for, a presumption for, whatever policies promise to alleviate this awful condition. To say that no policies have worked would be a simple error; everything, or almost everything, "works" to some extent or another. But we have been crushingly disappointed by all our policies.

Have goals and quotas helped? Perhaps. It was my argument in this book that they help those who do not need help and hardly reach those who do. I am not sure I am right. Complex econometric analyses come out with different stories. If we take as a measure the condition of the black lower class, certainly goals and quotas have been no marked success.

One can say the same for busing. Some studies show retrogression or stasis; some show progress in black achievement we might not have otherwise expected. But all this hardly makes a decisive success—or a success of any kind, when one throws into the balance the massive loss of white students, the sense of outrage among those affected, the costs and disruption inevitably present in a school system that must assign students by race and ethnic group.

But if the condition of lower class blacks provides the motor for the establishment and continuance of the quota society, and strengthens resistance to rolling it back, the machinery is a different story: It is provided by government regulatory agencies, by government officials, who may or may not "represent" these clients, but are part of the inexorable growth of governmental power outside the legislature and independent of elected legislators, or for that matter elected executives. The source of this power is endlessly debated and analyzed, but that it has grown enormously is unquestioned. Is it because Congress, out of incapacity or political calculation, refuses to legislate and review with the kind of detail, care, and specificity that would constrain government agencies and would put in place "the people's will" (as well as an elected body can represent it) rather than that of the permanent officialdom? Is it because the elected executive is incapable of controlling officials, whether politically appointed or protected by Civil Service, or again, refuses to take the political heat small minorities could bring if it did try to control them? I discuss this problem in chapter 6 of this book. But now there is more to say.

One of the most sobering developments in this entire story occurred after this book was published. President Gerald Ford and his cabinet officers (the Attorney General and the Secretary of Labor, and perhaps the Secretary of Health, Education, and Welfare) were doubtful of the constitutionality, wisdom, or efficacy of government regulations requiring employment and promotion by race and ethnic group, and of court judgments requiring school assignment of race. But they failed to take any effective action against these policies. They could not change the regulations of agencies they—in theory and law—controlled; they were constrained against entering litigation on school busing to oppose decisions they felt were wrong. Certainly they wanted to do so—or acted as if they did. Unquestionably they were supported by Congressional opinion. They were supported overwhelmingly by public opinion. But eight years of Republican administration saw these policies more decisively and effectively imposed on the country than anyone dreamed possible in 1968. When the Democrats returned in 1976, all the regulations remained in place for further development toward a society in which race and ethnic group determine individual fate. We now had top administrators who believed in this vision as well as the lower officialdom who have always tried to achieve it.

Why could not an administration that believed that this vision was wrong act? First, because a minority who did want such a society (I speak not of blacks and Hispanic Americans, who oppose a quota society almost as much as other Americans; I speak of those who *claim to* speak for them) had behind them the moral authority that comes from speaking for a group that has suffered fierce prejudice and discrimination and is now ridden with social problems. This moral authority was expressed in the way they could denounce those who wanted to moderate race and group based policies as racist, as wishing to undo the gains of the civil rights revolution and to return to 1876. But this authority was demonstrated more decisively in the degree to which these extreme and false charges were accepted by the national news media, by national commentators, by a large part of the body of professionally qualified opinion on these matters in the universities, government, the media. Whether or not we have a "new class" in this country—based on the universities, higher education, the media, and government—acting as self-appointed protectors of the poor and minorities, disdainful of the interests of those whom they see as "Middle Americans" without concern for the poor, what happened when the Ford administration tried to act to moderate goals, quotas, and school busing certainly suggests that a new source of political power now dominates decision making on race and ethnicity.

The shift in power from the private sector—business and trade unions and non-profit organizations—to government has been taking place for a long time. The decisive shifts in power are now *within* government, among Congress, the executive branch, the permanent officialdom, the regulatory agencies, and the courts, all involved in an ever shifting web of alliances and oppositions. Affirmative discrimination finds as its main sources of power within government the permanent officials of the agencies created to impose it and the regulatory agencies which move in the same direction. Resistance or assistance from the elected executive is not unimportant, but if President Ford was unable to hold the bureaucracy back, it may not matter that much that President Carter and Secretary Califano wish to spur it on. The courts, too, have provided important assistance, and occasional resistance. Congress has hardly mattered. It has remained quiescent, happy to take political advantage from criticizing the bureaucracy without taking the risk of denunciation by the "new class" by truly trying to control it. The courts, I believe, remain the only part of government capable of acting against the permanent bureaucracy, the regulatory agencies, and their allies. The arbitrary power which has enabled courts to forward the movement toward a quota society also permits them to take strong action against it, which is why this issue seems to have reached a critical point of decision in the Bakke case. But even the Supreme Court, because it wields such awesome power, is likely to hold back from radical action, leaving the main field of play in the hands of the government bureaucracies, regulatory agencies, and their "new class" allies.

The pattern of development has been different in the school busing issue. There the courts have played a more important role. The bureaucracy was less aggressive during a good part of the Nixon and Ford administrations, but this hardly mattered, as the courts were ready to go forward. It is an illusion that Nixon and Ford were able to make their opposition to busing effective within the bureaucracy. But certainly it is true that the bureaucracy was unleashed under the Carter administration. Here too, the courts, which have done so much to impose school assignment by race as a norm in American society, are the only force capable of stopping its expansion. Once again, it is doubtful that the Supreme Court will actually roll back what exists.

I have no intention of "predicting" the course of the Supreme Court. What is interesting and curious is that the struggle over affirmative discrimination, over the creation of a society that makes race and ethnic group the basis of official and private action, is now reduced to only two potential major antagonists: the Supreme Court on the one hand, the regulatory agencies and permanent bureaucracy on the other. It has, alas, become too dangerous for Congress or the President—or so, in their pusillanimity, they believe—to intervene decisively: They are content to let the shape of American society be determined by the Supreme Court and the permanent officialdom.

But that is hopefully not the whole story. There is still a role for business, trade unions, universities, and simply ordinary people. They can, and increasingly do, intervene in this struggle, opposing the officialdom through the courts. The courts thus become the final battleground, at least for this issue, and are forced into legislative and administrative roles. There is still a way to make one's voice felt, even if it is not by voting, but by litigation and *amicus* briefs, and it still offers hope that policies that most Americans reject as opposed to the animating vision of a good and just society may be stayed or reversed.

NATHAN GLAZER
March 20, 1978

ACKNOWLEDGMENTS

This book is based on the William W. Cook Lectures on American Institutions, which I delivered at the University of Michigan in April 1974, under the title "The American Ethnic Pattern: A New Phase?" I have revised, reorganized, and extended the lectures for publication in book form. In addition, I have added two chapters, not presented as separate lectures, based on previously published articles: Chapter 3, "Affirmative Action in Education: The Issue of Busing," draws heavily from "Is Busing Necessary?" (*Commentary,* 53 [March 1972]), and Chapter 4, "Overcoming Residential Segregation," draws heavily from "On 'Opening Up' the Suburbs" (*The Public Interest,* No. 37 [Fall 1974]). I am deeply grateful to Dean Theodore J. St. Antoine and the members of the committee in charge of the Cook Lectures for giving me the opportunity to present them, and the occasion that required me to fill out and develop my ideas on the subject. The work on the revision and extension of these lectures took place while I held a fellowship from the National Endowment for the Humanities, and I wish to express my gratitude to the Endowment. Martha Metzler typed and retyped this manuscript in its various stages with care and accuracy, and informed me of many places where I was obscure or incomplete; I am grateful to her.

NATHAN GLAZER

Cambridge, Massachusetts
May 1975

AFFIRMATIVE DISCRIMINATION: ETHNIC INEQUALITY AND PUBLIC POLICY

CHAPTER

1

The Emergence of an American Ethnic Pattern

IN THE MIDDLE of the last decade, we in the United States seemed to have reached a national consensus as to how we should respond to the reality of racial and ethnic-group prejudice and racial and ethnic-group difference. Almost simultaneously, we began to move away from that consensus into new divisions and a new period of conflict and controversy. The consensus was marked by three major pieces of legislation: the Civil Rights Act of 1964, the Voting Rights Act of 1965, and the Immigration Act of 1965. Following the passage of the Civil Rights and Voting Rights acts, the Federal government intervened firmly in the South to end the one hundred-year resistance of the white South to full political, civil, and social equality for blacks, insofar as this resistance

3

was embodied in law and public practice. The passage of the Immigration Act of 1965 marked the disappearance from Federal law of crucial distinctions on the basis of race and national origin. The nation agreed with this act that there would be no effort to control the future ethnic and racial character of the American population and rejected the claim that some racial and ethnic groups were more suited to be Americans than others.

In the phrase reiterated again and again in the Civil Rights Act of 1964, no distinction was to be made in the right to vote, in the provision of public services, the right to public employment, the right to public education, on the ground of "race, color, religion, or national origin." Paradoxically, we then began an extensive effort to record the race, color, and (some) national origins of just about every student and employee and recipient of government benefits or services in the nation; to require public and private employers to undertake action to benefit given groups; and school systems to assign their children on the basis of their race, color, and (some) national origins. This monumental restructuring of public policy to take into account the race, color, and national origin of individuals, it is argued by Federal administrators and courts, is required to enforce the laws against discrimination on these very grounds. It is a transitional period, they say, to that condition called for in the Constitution and the laws, when no account at all is to be taken of race, color, and national origin. But others see it as a direct contradiction of the Constitution and the laws, and of the consensus that emerged after long struggle in the middle 1960s.

I will examine in this book policies in three areas: employment, school desegregation, and residential location. I will analyze the position of those who support present policies and will argue that the consensus of the middle 1960s has been broken, and that it was and remains the right policy for the United States—right for the groups that had suffered, and

4

in some measure still suffer, from prejudice and discrimination, and right for the nation as a whole.

But the first step is to try to characterize and understand the consensus of the middle 1960s. This is not to be understood as an historically new response to the unprecedented events of those years—the vicious resistance in great parts of the South to the efforts of blacks to practice their political rights, the South's resistance to school desegregation, the shocking assassination of a President identified with the hopes of suppressed minority groups. It is to be understood rather as the culmination of the development of a distinctive American orientation to ethnic difference and diversity with a history of almost 200 years. That orientation was shaped by three decisions. They were not taken all at once, or absolutely, or in full consciousness of their implications, but the major tendencies of American thought and political action have regularly given their assent to them.

The three decisions were:

First, the entire world would be allowed to enter the United States. The claim that some nations or races were to be favored in entry over others was, for a while, accepted, but it was eventually rejected. And once having entered into the United States—and whether that entry was by means of forced enslavement, free immigration, or conquest—all citizens would have equal rights. No group would be considered subordinate to another.

Second, no separate ethnic group was to be allowed to establish an independent polity in the United States. This was to be a union of states and a nation of free individuals, not a nation of politically defined ethnic groups.

Third, no group, however, would be required to give up its group character and distinctiveness as the price of full entry into the American society and polity.

There is of course an inevitable breathtaking arrogance in asserting that *this* has been *the* course of American history.

It would be almost equally breathtaking to assert that *any* distinctive course can be discerned in the history of the shaping of the American people out of many different stocks. It is in part an act of faith to find any *one* course that the development of American society has in some way been reaching toward: It smacks of the unfashionable effort to give a "purpose," a direction, to history. Certainly this direction is not to be thought of as some unconscious immanent tendency continuing to reveal itself in American history. Direction in history is taken only in the concrete actions of men and of groups of men. Those actions in the United States have included —in direct conflict with the large direction I have described— the enslavement of the Negro, anti-immigrant and anti-Catholic movements that have arisen again and again in American life, the near extermination of the American Indian, the maintenance of blacks in a subordinated and degraded position for a hundred years after the Civil War, the lynching of Chinese, the exclusion of Oriental immigrants, the restriction of immigration from Southern and Eastern Europe, the relocation of the Japanese and the near confiscation of their property, the resistance to school desegregation, and so forth. If we are to seek a "direction" in American history that defines a distinctive approach to the relationship of the various ethnic groups that make up American society, the sequence of events just listed might well be made the central tendency of American history. Many current writers and scholars would have it so: They argue that racism defines our history—racism directed against blacks, Indians, Mexican Americans, Puerto Ricans, Filipinos, Chinese, Japanese, and some European ethnic groups. Many would have it that even the last ten years should be interpreted as a losing battle against this racism, now evident in the fact that colleges and universities resist goals and targets for minority hiring, that preferential admissions to professional schools are fought in the courts, that the attempt to desegregate the schools in the North and West has now met a resistance extremely difficult to overcome, that housing for poor and minority groups is excluded from many suburbs.

I think this is a selective misreading of American history: that the American polity has instead been defined by a steady expansion of the definition of those who may be included in it to the point where it now includes all humanity; that the United States has become the first great nation that defines itself not in terms of ethnic origin but in terms of adherence to common rules of citizenship; that no one is now excluded from the broadest access to what the society makes possible; and that this access is combined with a considerable concern for whatever is necessary to maintain group identity and loyalty. This has not been an easy course to shape or maintain, or to understand. There have been many threats to this complex and distinctive pattern for the accommodation of group difference that has developed in American society. The chief threats in the past were, on the one hand, the danger of a permanent subordination of certain racial and ethnic groups to others, of the establishment of a caste system in the United States; and on the other hand, the demand that those accepted into American society become Americanized or assimilated, and lose any distinctive group identity. The threat of the last ten years to this distinctive American pattern, however, has been of quite another sort. The new threat that followed the most decisive public actions ever taken to overcome subordination and caste status was that the nation would, under the pressure of those recently subordinated to inferior status, be permanently sectioned on the basis of group membership and identification, and that an experiment in a new way of reconciling a national polity with group distinctiveness would have to be abandoned. Many did not and do not see this latter possibility as any threat at all, but consider it only the guarantee and fulfillment of the commitment of American society to admit all peoples into full citizenship. They see the threat to a decent multigroup society rising from quite another direction: the arrogance and anger of the American people, specifically those who are descended from colonists and earlier immigrants, aroused by the effort to achieve full equality for all individuals and all groups. The prevailing understanding of the present mood is that those who have their share—and more—want to turn their backs on the process that is necessary to dismantle a caste society in which some groups are held in permanent subordination. I think this is a radical misreading of the past few years, and, in subsequent chapters, I will argue why.

But the task of this first chapter is a rather different one: If the history of American society in relationship to many of the groups that make it up is not a history of racism, what is it? How do we define an emergent American attitude toward the problem of the creation of one nation out of many peoples?

I have suggested there were three major decisions—decisions not taken at any single point of time, but taken again and again throughout our history—which defined this American distinctiveness. The first was that all should be welcome and that the definition of America should be a political one, defined by commitment to ideals, and by adherence to a newly created or freshly joined community defined by its ideals, rather than by ethnicity. Inevitably "American" did come to denote an "ethnicity," a "culture," something akin to other nations. A common life did create a common culture, habits, language, a commonness which parallels the commonness of other nations with their more primordial sense of commonness. But whereas all European and many Asian nations have grown out of a primordial group of long history, bound together by culture, religion, language, in the American case there was a continual struggle between the nation understood in these terms—terms akin to those in which the French, or English, or Germans understood themselves—and the nation understood in very different terms.

Yehoshua Arieli describes a number of ways in which a pattern of national identification has been achieved. In some cases, national identification is imposed by force; in others, it has grown gradually, and ". . . resulted from a long-established community of life, traditions, and institutions. . . ." But ". . . in the United States, national consciousness was shaped by social and political values which claimed universal validity and which were nevertheless the American way of life. Unlike other Western nations, America claimed to possess a 'social system' fundamentally opposed to and a real alternative to . . ."—and here I edit Arieli, but not against his meaning—". . . [other social systems], with which it competed by claiming to represent the way to ultimate progress and a true social happiness." [1] These different terms in which American nationality defined itself consisted not only of the decisive revolutionary act of separation from England and self-definition as a separate nation. They also included for many of those who founded and helped define the nation the rejection of ethnic exclusivity.

Three writers to my mind have, in recent years, given the best definition of what it meant to found a nation in this way: Seymour Martin Lipset in *The First New Nation;* Hans Kohn in his book *American Nationalism: An Interpretive Essay;* and Yehoshua Arieli in *Individualism and Nationalism in American Ideology.*

Arieli argues forcefully that the American Revolution should *not* be seen as another uprising of an oppressed nation, but as an event whose main shapers presented it as significant for the world and all its peoples:

All the attempts made by Americans to define the meaning of their independence and their Revolution showed an awareness that these signified more than a change in the form of government and nationality. Madison spoke of the American government as one which has "no model on the face of the globe." For Washington, the United States exhibited perhaps the first example of government erected on the simple principles of nature, and its establishment he considered as an era in human history. . . . John Adams was convinced that a greater question than that of American Independence "will never be decided among men."

For Jefferson, America was the proof that under a form of government in accordance "with the rights of mankind," self-government would close the circle of human felicity and open a "widespread field for the blessings of freedom and equal laws." Thomas Paine hailed the American Revolution as the beginning of the universal reformation of mankind and its society with the result "that man becomes what he ought." For Emerson, America was " . . . a last effort of the Divine Providence in behalf of the Human race." [2]

We might of course expect a second-generation sociologist, a scholar who found refuge here, and another refugee who has become a scholar in a newly founded democratic nation to respond to these claims, to reverberate to them, so to speak. We might also expect Jewish scholars to respond to these claims, for if the United States was very late in fulfilling its promise to blacks, Indians, Mexican Americans, and others—that is, those of other races—it almost from the beginning offered an open field and freedom to those who practiced another religion. Can a more searching examination, however, sustain these claims? Could it not also be said that American independence and the establishment of a new country was little more than the assertion of the arrogance of British colonists, refusing to accept a moderate overseas government more solicitous of the rights of Indians and blacks than they were, insisting on taking the land from the Indians and on the right to import and hold black men as slaves, and eventually threatening their neighbors with imperial expansion?

Our history, indeed, included all this. Even the appropriation of the name "American" for the citizens of the United States is seen by our neighbors to the north and to the south as a symbol of arrogance. Yet other interpretations of this appropriation are possible. The Americans *did* accept as the name for themselves a name with no ethnic reference, a name even with no limited geographical reference (since the Americas include all the Western Hemisphere). One side of this self-naming may be seen as a threat to the rest of the Americas and as arrogance in ignoring their existence. But another side must also be noted: the rejection by this naming of any reference to English or British or any other ethnic or racial origins, thus emphasizing in the name itself the openness of the society to all, the fact that it was not limited to one ethnic group, one language, one religion.

Lipset argues that the American nation from the beginning established and defined its national identity on the basis of its decisive break, through revolution, with England, and, by extension, with the entire old world. This weakened the ethnic identification with England. Further, two values became dominant in American society and the shaping of American character, equality and achievement, and these values can be seen sharply marked in American society from the beginning of its independent political existence.[3] One point about these two values that I would emphasize is that, by their nature, they cannot remain ethnically exclusive. And the most far-sighted of the early leaders understood this. Thus, to quote Hans Kohn:

Thomas Jefferson, who as a young man had opposed immigration, wished in 1817 to keep the doors of America open, "to consecrate a sanctuary for those whom the misrule of Europe may compel to seek happiness in other climes." . . . This proclamation of an open port for immigrants was in keeping with Jefferson's faith in America's national mission as mankind's vanguard in the fight for individual liberty, the embodiment of the rational and humanitarian ideals of eighteenth century man.
    The American nation,

Hans Kohn continues, summarizing Jefferson's point of view,

was to be a universal nation—not only in the sense that the idea which it pursued was believed to be universal and valid for the whole of mankind, but also in the sense that it was a nation composed of many ethnic strains. Such a nation, held together by liberty and diversity, had to be firmly integrated around allegiance to the American idea, an idea to which everyone could be assimilated for the very reason that it was a universal idea. To facilitate the process of integration, Jefferson strongly opposed the settlement of immigrants in compact groups, and advocated their wide distribution among the older settlers for the purpose of "quicker amalgamation." [4]

Of course, to one tradition we can oppose another. If Jefferson was positive about the immigration of other groups, Benjamin Franklin was suspicious. "For many years," Arieli writes, "he strenuously argued against the wisdom of permitting the immigration of non-English settlers, who 'will never adopt our language or customs anymore than they can acquire our complexion.' " [5] Undoubtedly, he was influenced by the substantial number of Germans in Pennsylvania, itself established as an open colony of refuge:

This will in a few years [Franklin wrote] become a German colony: Instead of their Learning our Language, we must learn theirs, or live as in a foreign country. Already the English begin to quit particular Neighborhoods surrounded by Dutch, being made uneasy by the Disagreeableness of dissonant Manners; and in Time, Numbers will probably quit the Province for the same Reason. Besides, the Dutch under-live, and are thereby enabled to under-work and under-sell the English; who are thereby extremely incommoded, and consequently disgusted, so that there can be no cordial Affection or Unity between the two Nations. [6]

The themes are, of course, familiar ones: They were to be repeated for many groups more distant from the Anglo-American stock than the Germans, who were, after all, of related tongue and Protestant religion. And yet this was a private comment, to be set against a public one that, again to quote Kohn, "extolled Anglo-America as a place of refuge." [7]

There were two traditions from the beginning, traditions exemplified by different men and social groups, and carried in tension within the same men. Yet even to say there were two traditions makes the issue somewhat sharper than it could have been during the early history of the United States. After all, the very men who spoke about the equal rights of all men accepted slavery. If they spoke of the United States as a sanctuary for all, they clearly thought of men very like themselves who might be seeking it and were not confronted with the hard realities of men of very different culture, religion, and race taking up their offer. In addition, we must take account of the expansive rhetoric of a moment in which a nation was being founded. Yet stipulating all of these cautions, there was a development implied in the founding documents and ideas which steadily encouraged the more inclusive definitions of who was eligible to become a full participant in American life. In the Revolution and its aftermath, limitations on participation in public life by the propertyless, Catholics, and Jews were lifted. Waiting in the wings, so to speak, were other categories, implied in the founding principles. That some others waited for almost two centuries, and that their equality came not only because of founding principles but because of complex social and political developments is true; but the principles were there, exerting their steady pressure, and indeed even in 1975 much of the argument over how to define full equality for different groups revolves around a Constitution that dates to 1787.[8]

As Arieli puts it: "Whatever the impact of universal concepts on the American historical experience, the conservative and nativistic interpreters of American history, no less than their opponents, concede that American nationality has to be defined, at least to some degree, by reference to certain political and social concepts; that it is a way of life and an attitude which somehow represents ultimate social values. . . ."[9]

There is no Supreme Historian, sitting in heaven, who totes up the record and tells us which way the balance of history ran. One picks out a dominant theme, on the basis of one's experience as well as one's knowledge, and our choice is made, in part, on the basis of our hopes for the future as well as our experience. In the 1950s and 1960s men like Kohn and Arieli wanted to emphasize the inclusive tradition; in the later 1960s and in the 1970s, many historians and other scholars

13

want to show us the exclusive tradition. There is enough to choose from on both sides. We can quote Melville, writing in 1849:

There is something in the contemplation of the mode in which America has been settled, that, in a noble breast, should forever extinguish the prejudices of national dislike. Settled by the people of all nations, all nations may claim her for their own. You cannot spill a drop of American blood without spilling the blood of the whole world. . . . We are not a narrow tribe of men with a bigoted Hebrew nationality—whose blood has been debased in the attempt to ennoble it, by maintaining an exclusive succession among ourselves. No; our blood is as the flood of the Amazon, made up of a thousand noble currents all pouring into one. We are not a nation, so much as a world. . . . On this Western Hemisphere all tribes and peoples are forming into one federal whole. . . .[10]

Or James Fenimore Cooper, writing in 1838:

The great immigration of foreigners into the country, and the practice of remaining, or of assembling, in the large towns, renders universal suffrage doubly oppressive to the citizens of the latter. The natives of other countries bring with them the prejudices of another . . . state of society; . . . and it is a painful and humiliating fact, that several of the principal places of this country are, virtually, under the control of this class, who have few convictions of liberty. . . . Many of them cannot even speak the language of the land, and perhaps a majority of them cannot read the great social compact, by which society is held together.[11]

In certain periods, it seems clear, one voice or another was dominant. The uprising of the white South in the Civil War marked the most determined effort to change the pattern into one in which other races and groups, labeled inferior, were to be held in permanent subjection and subordination. A new justification was to be established for this "heresy," as Arieli dubs it—and in the American context, heresy it was. Justification was to be found in religion, in pragmatic necessity, in political theory, even surprisingly enough in Auguste Comte's new-founded science of sociology, which was drawn upon to show the superiority of slave labor to Northern, immigrant, free labor, and of a society founded on slavery to one founded on free immigration.[12]

It is revealing that one great effort to avoid the conflict consisted of the rapid upsurge of the "American" party [the "Know-Nothings"] which labored to unite discordant political factions by making ethnic and religious loyalties the basis of national identification. It sought to substitute for traditional American values a nationalism of the Old World type based on common descent and religion, and thus to divert against the "foreigners" the antagonisms that existed among the native-born. Similarly, the theory of race which justified Negro slavery also aimed to create an identity between North and South on the basis of a common belief in white superiority and through territorial expansion. Yet the historical situation and the national tradition frustrated these efforts and turned the conflict between free and slaveholding states into a gigantic struggle over the nature of American social ideals.[13]

After early remarkable successes, the Know-Nothings disintegrated before the rise of the new Republican party, thus setting a pattern that other nativist movements were to follow again and again, such as the American Protective Association of the 1890s and the Ku Klux Klan of the 1920s—first a sudden upsurge that seemed to carry all before it, and then, equally suddenly, disintegration. The challenges to the central American pattern, brief and intense, were rapidly overtaken by the major tendency to a greater inclusiveness. The Know-Nothings disintegrated, and the South lost the war. The heresy was, for a while, extirpated.

In the wake of the Civil War, the great Southern heresy that had threatened the idea of American nationality as broadly inclusive seemed crushed. As John Higham writes of those postwar years:

America had developed a fluid, variegated culture by incorporating alien peoples into its midst, and the experience had fixed in American thought a faith in the nation's capacity for assimilation. This faith, carrying with it a sense of the foreigner's essential identification with American life, expressed itself in a type of nationalism that had long offset and outweighed the defensive spirit of nativism. A cosmopolitan and democratic ideal of nationality made assimilation plausible to Americans. . . .

The twin ideals of a common humanity and of equal rights continued in the 1870's and 1880's to foster faith in assimilation. Temporarily the tasks of post-war reconstruction even widened assimilationist ideals; for the Radical Republicans' effort to redeem the southern Negro, to draw him within the pale of the state, and to weld the two races into a homogeneous nationality discouraged emphasis on human differences. To James Russell Lowell, for example, just and equal treatment of black men meant simply an enlargement of the Christian mission which the United States had long performed in bringing together the peoples of all nations in a common manhood. And Elisha Mulford, philosopher of Reconstruction, argued that the nation "is inclusive of the whole people. . . . There is no difference of wealth, or race, or physical condition, that can be made the ground of exclusion from it." [14]

But of course, new threats and new heresies were rising, and the United States was soon to enter a dark age in which the promise of an all-embracing citizenship and nationality, already a hundred years old, was, for a while, quite submerged. Indeed, the very New England elite who had refused to accept slavery and celebrated the open door themselves began to undergo a significant change as the flood of immigration poured into the country after the Civil War, a flood that became increasingly Catholic, increasingly non-English speaking, increasingly Jewish and Central and Eastern European as the century wore on. By the 1890s, a new criticism—which took many forms—of an inclusive idea of American citizenship was arising. The New England intellectuals, now displaced politically and culturally, no longer carried on the tradition of the American revolution. Having attacked the racist ideology of the South before the Civil War, many succumbed to a new, if milder, racism which placed the Anglo-Saxon or "Germanic" element of the American people at the apex of world evolution as the carriers of some special racial commitment to liberty and free government. On quite a different cultural level, waves of anti-Catholicism spread through the masses of white Protestant farmers and workers, peaking in the American Protective Association of the late 1880s and 1890s, which had as rapid a rise—and fall—as Know-Nothingism before the Civil War. Anti-Semitism for the first time appeared in the United States. Some scholars discern it in the Populist movement of the 1890s, some do not. But all recognize it in an increasing exclusivism of wealthy Eastern Americans in the same period. In the West, an anti-Chinese movement became virulent among white workingmen, and led to the first restriction of American immigration in 1882. By the end of the decade, the strongest and darkest thread of this skein of prejudice and discrimination dominated the rest as the modest gains of Reconstruction were swept aside in the South, Jim Crow laws were fastened on the free Negro, the last Negro representatives were swept from Congress, and a rigid caste system was imposed upon the black, by law in the South and custom in the North.[15]

Each thread of this complex pattern deserves full analysis, and in the present mood, in which the American past is being reviewed by scholars representing many oppressed groups, and in a time when many see the United States as the chief force of reaction in the world, each thread is receiving such analysis. For fifty years, between the 1890s and the 1930s, exclusivism was dominant. It affected many groups—blacks and Orientals, Jews and Catholics, Indians and Mexican Americans—in many ways. One can at least explain some of the reasons for the reaction against admitting all people into the country and to full citizenship. People of position felt threatened by the incoming flood of immigrants. Workers and shopkeepers without stable positions also felt threatened, by the Chinese, the blacks, the Catholics, the immigrants, with the same fears that Franklin had expressed 150 years before over the Germans "under-living, under-selling, and under-working" the English. Those not in direct competition with the immigrant and the black felt the fears just as strongly, as we see in the case of the farmers, who tried to understand the sudden falls in price which threatened to destroy them by resorting to a belief in dark plots by international financial forces. Fears do not justify prejudice, discrimination, and racism, but they help explain it. And the expansion of American society to include strangers from all over the world was not without its real losses as well as its imaginary fears.

Barbara Solomon has recorded the story of the New England intellectuals at the turn of the century. Earlier, they had supported free immigration as well as abolition. Thus, long before Emma Lazarus, James Russell Lowell delivered some verses quite reminiscent of her poem on the Statue of Liberty, in an ode delivered at Harvard in 1861. He spoke of the American nation as:

> She that lifts up the mankind of the poor,
> She of the open soul and open door,
> With room about her hearth for all mankind! [16]

And in an address delivered in 1878, Emerson wrote: "Opportunity of civil rights, of education, of personal power, and not less of wealth; doors wide open . . . invitation to every nation, to every race and skin, . . . hospitality of fair field and equal laws to all. Let them compete, and success to the strongest, the wisest, and the best." [17] By the end of the century, many had quite given up their faith in democracy and the equality of all peoples and had become enamored of the notion that American liberty sprang from German forests and could not be maintained by the flood of immigrants from Eastern and Southern Europe. (Of course, not all New England intellectuals took this course. Preeminently in opposition were Charles Eliot, President of Harvard—though not, alas, his successor, President Lowell—and William James. Eliot wrote in 1920: "I should like to be saved from loss of faith in democracy as I grow old and foolish. I should be very sorry to wind up as the three Adamses did. I shall not, unless I lose my mind." [18]) Essentially, what troubled the New England intellectuals who no longer followed the democratic faith of Emerson was the threat to American homogeneity, for a measure of homogeneity had indeed existed before the heavier floods of immigration had begun. In our present-day mood of easy analysis of American racism, we would argue that they were defending their economic, political, and social interests. But their economic interests were not threatened by immigration: Quite the contrary, the immigrants gave New England industry an important source of cheap labor. Their political interests were not threatened, for their local political domination had already been lost to the Irish immigrants when the flood of East Europeans and Italians began seriously to concern them in the 1890s. Indeed, these new immigrants offered them, perhaps, a chance to regain political power. Their social interests were not deeply involved, for there seemed little chance that the new immigrants would join them in polite society.

Where, then, did they feel threatened? They felt they were losing their country, that what they knew of as America was disappearing and becoming something else, and that American culture was going to be radically changed into something they would not recognize. Small-town life, country pleasures, certain forms of education, modes of recreation, characteristic tendencies in religion—this whole complex, they feared, was in danger. As we learn from Solomon's book, no crude or simple prejudice activated them. Many of the old New Englanders who favored immigration restriction were active in social work among the immigrants, and some were patrons of bright immigrant youths. But they did not want to see the American culture they knew go.

As one of them wrote, in a passage quoted by Horace M. Kallen: "We are submerged beneath a conquest so complete that the very name of us means something not ourselves. . . . I feel as I should think an Indian might feel, in the face of ourselves that were." [19] Henry James, returning to this country in 1907 after an absence of twenty-five years—he still considered himself an American—expressed this shock most vividly: "This sense of dispossession . . . haunted me so, I was to feel . . . that the art of beguiling or duping it became an art to be cultivated—though the fond alternative vision was never long to be obscured . . . of the luxury of some such close and sweet and *whole* national consciousness as that of the Switzer and the Scot." [20]

Toward the end of his life, William Dean Howells, who had enjoyed seeing French Canadians and Italians around Boston in the 1870s and had praised the Jewish immigrant writer Abraham Cahan, wrote a novel, *The Vacation of the Kelwyns,* subtitled *An Idyll of the Mid-1870's.* We are introduced to a New England landscape in the year of the centennial celebrations of 1876. Kelwyn, a university lecturer, is spending the summer with his family in a large Shaker family house, now empty. The New England countryside is slightly menacing that year with "tramps"—unemployed workers—and foreigners wander through it: a Frenchman with a trained bear, Italian organ grinders, some gypsies. Kelwyn bears no antipathy to them—quite the contrary, they enliven the scene, and, as he says, they cook better than the natives, and may help make life pleasanter. And yet, one feels an unutterable sadness over the passing of a peculiarly American civilization. The Shaker house is empty, and will never be filled again; the surviving Shakers have misguidedly furnished it for the Kelwyns with new furniture; the New England countryside is different from what it has been; and the year of the centennial —Howells seems to be saying—marks the passing of something simple and sweet in America. Indeed, the centennial itself involves deeper feeling in some of the characters in the book than we can imagine any national celebration since evoking.

In the North, exclusivism expressed itself in resistance to

immigration from Eastern and Southern Europe and suspicion of immigrant settlements in the cities—of their habits, their culture, their impact on political life and on urban amenities. The Negroes were present—they always had been—but they were so few and so far down the social scale that they were scarcely seen as a threat to anything. In the South, exclusivism was directed primarily against the Negroes, though Catholics and Jews came in for their share of prejudice and, on occasion, violence. In the West, the Chinese and the Japanese were the main targets of a pervasive racism which included the Mexicans and the Indians.

The dismantling of this system of prejudice and discrimination in law and custom began in the 1930s. In the North, the ethnic groups created by the new immigration began to play a significant role in politics; and blacks, after the disenfranchisement of the 1890s, began again to appear in politics. The last mass anti-Catholic movement was the Klan's in the 1920s. It had a short life, and was in eclipse by the time Al Smith ran for President in 1928. Anti-Semitism had a longer life, but the war against Hitler ended with the surprising discovery that anti-Semitism, so strong in the Thirties, was undergoing a rapid and unexpected deflation. And similarly with anti-Chinese and Japanese prejudice. The immigration restriction law of 1924 was modified to accept at least token numbers of people from all nations and races in 1952, and all elements of national or racial preference were expunged in 1965.

Of course, the major bastion of race discrimination was the South, and the legal subordination of the Negro there remained firm throughout the 1930s and 1940s. But twenty years of liberal domination of national politics, by a coalition in which in Northern cities blacks played an important role, finally made its effects felt in the administration of President Truman. The Armed Forces were desegregated, national demands for the enfranchisement of Southern blacks became stronger and began to receive the support of court decisions, and a major stage in the elimination of discriminatory legislation was reached with the Supreme Court decision of 1954 barring segregation in the public schools. With the Civil Rights Act of 1964 and the Voting Rights Act of 1965, the caste system of the South was dismantled. The thrust for equality now shifted from the legal position of the group to the achievement of concrete advances in economic and political strength.

Thus for the past forty years, the pattern of American political development has been to ever widen the circle of those eligible for inclusion in the American polity with full access to political rights. The circle now embraces—as premature hyperbolic statements made as long as 200 years ago suggested it would—all humanity, without tests of race, color, national origin, religion, or language. To what extent an equalization of economic position has been associated with this political equalization is discussed in the second chapter.

Two other elements describe the American ethnic pattern, and these are not as easily marked by the processes of political decision-making, whether by court, legislature, or war. The first additional element is that the process of inclusion set limits on the extent to which different national polities could be set up on American soil. By "polity" I refer to some degree of political identity, formally recognized by public authority. Many multiethnic societies do recognize different ethnic groups as political entities. In the Soviet Union, each is formally entitled to a separate state or autonomous region (though these distinctive units exercise their powers in a state in which all individuals and subunits are rigidly controlled by a central dictatorship). Even a group dispersed throughout the Soviet Union such as the Jews is recognized as a separate nationality; and at one time, this nationality had rights, such as separate schools, publications, publishing houses. In Eastern Europe, where successor states to the German, Russian, and Austro-Hungarian Empire were set up after World War I, once again national rights were given to groups, even to such groups as the Jews, who were dispersed throughout the national territory. In nations that have been created by migration, such as the United States, we do not have examples of something like "national rights."

But the United States is more strict than others in preventing the possibility that subnational entities will arise. Consider the case of Canada, which is also a multiethnic society. The major minority national group, the French, is a compactly settled group which was conquered in the eighteenth century: It was not created through migration into a preexisting homogeneous or multiethnic nation. There are far more extensive national rights for the French than the United States allows for any group. Bilingualism is recognized not only in the areas of French settlement, but throughout the country. It is required of civil servants.

But what is the position of the "third element," those neither of English nor French origin? Their language has rapidly become English (except in the isolated prairie settlements of Eastern European farmers); nevertheless, as French rights have become more and more secured, they, too, have come to demand some special rights. In contrast with the United States, Canadian nationality is made up of *two* distinct founding ethnic groups, the French Canadians and the descendants of settlers from the original conquering power, to which have been added many ethnic groups derived from immigrants. There is a certain resistance among the third element—whether Jews in Montreal and Toronto, or Ukrainians in the West—to identifying fully with or assimilating to one or the other founding group, in part because these founding groups persist in maintaining specific ethnic characteristics of an English or French character. This is understandable: Canada was not founded in a revolutionary break from the fatherland, whether French or English, and while it is true a distinct Canadian personality and character did develop in both the French and

English element, no great emphasis was placed on specifically distinguishing "Canadian" from everything "non-Canadian." [21] This created a problem for new immigrant groups: Were they to maintain their original ethnic characteristics to the same extent that the founding groups did? And it created a problem for the polity. To what extent were the new immigrant groups to be encouraged to do so, or hindered in doing so? Thus, becoming Canadian did not imply, to the same extent that becoming American did, an abandonment of immigrant ethnic traits and a becoming something different. And so, the assimilation of ethnic groups in Canada did not proceed as rapidly as that of their ethnic relatives in the United States.

Among the possibilities for making political accommodation to groups of different ethnic character in a contemporary state, the United States falls near one end of the spectrum in denying formal recognition for any purpose to ethnic entities. In contrast to Canada, we do not ask for "ethnicity" in our census—though some government census sample surveys have recently done so—nor do we demand that each respondent select an ethnic origin.

Our pattern has been to resist the creation of formal political entities with ethnic characteristics. The pattern was set as early as the 1820s, when, as the historian of American immigration, Marcus Hansen, describes it, a number of European groups thought of establishing a New Germany or a New Ireland in the United States. He writes:

The first step in any of these dreams [to establish branches of other nations on American soil] was the acquisition of land. But the government of the United States, though possessed of millions of acres, proved unwilling to give a single acre for the purpose. It expressed its opinion in unmistakeable terms in the year 1818 when the Irish societies of New York and Philadelphia, burdened with a large number of charitable cases, petitioned Congress for a land grant in the West on which to establish their dependents. Congress refused, agreeing with the report of a special committee that it would be undesirable to concentrate alien peoples geographically. If a grant were made to the Irish, the Germans would be the next, and so with other nationalities. The result would be a patchwork nation of foreign settlements. Probably no decision in the history of American immigration policy possesses more profound significance. By its terms the immigrant was to enjoy no special privileges to encourage his coming; also he was to suffer no special restrictions. His opportunities were those of the native, nothing more, nothing less.[22]

No new nations would be established on American soil. We were to be, if a Federal republic, a republic of states, and even the states were not to be the carriers of an ethnic or national pattern. Most divergent from this norm, perhaps, is New Mexico, a state created out of conquered territory with a settled population, or the special rights of the Spanish-origin settlers of the Gadsden Purchase in Arizona; but even in those states, the rights of the Spanish-speaking barely lead to the creation of an ethnic state, although some militant Chicano leaders would perhaps like to see this happen.

Finally, there was a third set of decisions that defined the American ethnic pattern: Any ethnic group could maintain itself, if it so wished, on a *voluntary* basis. It would not be hampered in maintaining its distinctive religion, in publishing newspapers or books in its own language, in establishing its own schools, and, indeed, in maintaining loyalty to its old country.

This was a policy, if one will, of "salutary neglect." If immigrants could not establish new polities, they could do just about anything else. They could establish schools in their own language. They could teach their own religion, whether it was the ancient faith of Rome or the newly founded variants of Judaism and Islam developed by American blacks. When the state of Washington tried, in the early 1920s, to make public education a state monopoly, the Supreme Court said it could not.[23] Immigrants could establish their own churches and, under the doctrine of state-church separation, these would neither be more favored nor less favored than the churches of the original settlers which had once been established churches. They could establish their own hospitals, cemeteries, social service agencies to their own taste. All would be tax exempt: The state, in effect, respected whatever any group more or less wanted to consider education, or health and welfare, or religion, or charity. (Polygamy was one exception.) Indeed, the hospitals and social service agencies of these groups were even eligible for state funds, just as the institutions set up by the churches and groups of the early settlers had been. Immigrants could send money freely to their homelands, they could support the national movements of their various groups, and they could also, relatively easily, get tax exemption for their contributions to anything that smacked of religion, education, health and welfare, or charity.

There was no central public policy organized around the idea that the ethnic groups were a positive good, and therefore should be allowed whatever freedom they needed to maintain themselves. Policymakers generally never thought of the matter. It was, rather, that there was a *general* freedom, greater than in most other countries, to do what one willed. The mere fact that city planning and the controls associated with it were so much weaker than in other countries made it easy to set up churches, schools, and the like. In a society in which land could easily be bought and sold, fortunes easily made (and unmade), and mobility was high, there were, in effect, two sets of forces set loose: One force tended to break up the ethnic communities, for it was easy in American society to distance oneself from family and ethnic group, if one wanted to; but at the same time, and this is what is often forgotten, it was also easy to establish the institutions that one desired. This meant, of course, that every church divided again and again: The state was disinterested, and thus every variant of liberalism and orthodoxy could express itself freely in institutional form. It also meant there was no hindrance to the maintenance of what one wished to maintain.

One of the interesting general findings of ethnic research is that affluence and assimilation have double effects. On the one hand, many individuals become distant from their origins, throw themselves with enthusiasm into becoming full "Americans," and change name, language, and religion to forms that are more typical of earlier settlers. On the other hand, however, many use their increased wealth and competence in English to *strengthen* the ethnic group and its associations. It is hard to draw up a balance as to which tendency is stronger, because different people evaluate different effects differently. Thus undoubtedly, with longer residence in the United States, folk aspects of the culture weaken, and those attached to them feel that the original culture is lost. Yet associational and organizational forms of ethnicity are strengthened. For example, the one-room school, the *heder,* where Jewish children learned their letters, their prayers, and a bit of Bible under the tutelage of an Old World teacher, disappeared; so it became possible to say that the true old East European Jewish culture was gone. But regularly organized religious and Hebrew schools, with classrooms and teachers after the American pattern, increased greatly in number, and more Jewish children had some formal Jewish education under the organized system than under the folk system. Or, to take another example, undoubtedly, in 1975, the more folkish aspects of Ukrainian culture have weakened, both for the pre-World War II and post-World War II immigrants. This weakening is associated with assimilation and higher income. But now there are chairs for Ukrainian studies at Harvard, supported by funds raised by Ukrainian students. It is this kind of trade-off that makes it so difficult to decide whether there is really, as Marcus Hansen suggested, a third-generation return to ethnic origins and interests. There is a return, but as is true of any return, it is to something quite different from what was there before.

In any case, whatever the character of the return, it is American freedom which makes it possible, as American freedom makes possible the maintenance and continuity and branchings out of whatever part of their ethnic heritage immigrants and their children want to pursue.

When we look now at our three sets of decisions—that all may be included in the nation, that they may not establish new nations here, and that they may, nevertheless, freely maintain whatever aspects of a national existence they are inclined to—we seem to have a classic Hegelian series of thesis, antithesis, and synthesis. The synthesis raises its own new questions, and these become steadily more sharp, to the point where many argue we must begin again with a new thesis. For the three sets of decisions create an ambiguous status for any ethnic group. The combination of first, you may become full citizens; second, you may not establish a national entity; third, you may establish most of the elements of a national entity voluntarily without hindrance, does not create an easily definable status for the ethnic group. The ethnic group is one of the building blocks of American society, politics, and economy, none of which can be fully understood without reference to ethnic group formation and maintenance, but this type of group is not given any political recognition or formal status. No one is "enrolled" in an ethnic group, except American Indians, for whom we still maintain a formally distinct political status defined by birth (but any individual Indian can give up this status). For all public purposes, everyone else is only a citizen. No one may be denied the right to political participation, to education, to jobs because of an ethnic status, nor may anyone be given better access to political appointments or election, or to jobs, or education because of ethnic status. And yet we pore over the statistics and try to estimate relative standings and movements among ethnic groups.

A distinction of great importance to our society is thus given no formal recognition and yet has great meaning in determining the individual's fate. In this sense, ethnicity is akin to "class" in a liberal society. Class does not denote any formal status in law and yet plays a great role in the life of the individual. Ethnicity shares with class—since neither has any formal public status—a vagueness of boundaries and limits and uncertainty as to the degree to which any person is associated with any grouping. No member of the upper, middle, or lower class—or choose what terms you wish—needs to act the way most other members of that class do; nothing but social pressure will hold him to any behavior. Similarly with persons whom we would consider "belonging" to ethnic groups: They may accept that belonging or reject it. Admittedly, there are some groups, marked by race, where belonging is just about imposed by the outside world, as against other less sharply marked groups. Nevertheless, the voluntary character of ethnicity is what makes it so distinctive in the American setting. It is voluntary not only in the sense that no one may be required to be part of a group and share its corporate concerns and activities; no one is impelled *not* to be part of a group, either. Ethnicity in the United States, then, is part of the burden of freedom of all modern men who must choose what they are to be. In the United States, one is required neither to put on ethnicity nor to take it off. Certainly this contributes to our confusion and uncertainty in talking about it.

Undoubtedly, if this nation had chosen—as others have—either one of the two conflicting ideals that have been placed before us at different times, the "melting pot" or "cultural pluralism," the ambiguities of ethnic identity in the United States and the tensions it creates would be less. Under the first circumstance, we would have chosen a full assimilation to a new identity. Many nations have attempted this: some forcefully and unsuccessfully, as did Czarist Russia in relation to certain minority groups; some with a supreme self-confidence, such as France, which took it for granted that the status of the French citizen, *tout court,* should satisfy any civilized man; some, with hardly any great self-consciousness, such as Argentina, which assimilated enormous numbers of European immigrants into a new identity, one in which they seemed quite content to give up an earlier ethnic identity, such as Spanish or Italian. If a nation does choose this path of full assimilation, a clear course is set before the immigrant and his children. Similarly, if the principle is to be that of cultural pluralism, another clear course is set. We have not set either course, neither the one of eliminating all signs of ethnic identity—through force or through the attractions of assimilation—nor the other of providing the facilities for the maintenance of ethnic identity.

But our difficulties do not arise simply because of the ambiguities of personal identity. They arise because of the concrete reality that, even in a time of political equality (or as close to political equality as formal measures can ensure), ever greater attention is paid to social and economic inequality.

If we search earlier discussions of the immigrant and of ethnic groups, we will not find any sharp attention to these inequalities. It was assumed that time alone would reduce them, or that the satisfactions of political equality would be sufficient. It was assumed, perhaps, that social and economic inequalities would be seen as *individual* deprivations, not as *group* deprivations. But there was one great group whose degree of deprivation was so severe that it was clearly to be ascribed to the group's, not the individual's, status. This was the Negro group. As we concentrated our attention in the 1960s on the gaps that separated Negroes from others, other groups of somewhat similar social and economic status began to draw attention to *their* situation. And as these new groups came onto the horizon of public attention, still others which had not been known previously for their self-consciousness or organization in raising forceful demands and drawing attention to their situation entered the process. What began as an effort to redress the inequality of the Negro turned into an effort to redress the inequality of all deprived groups.

But how is this to be done? And does not the effort to redress upset the basic American ethnic pattern? To redress inequalities means, first of all, to define them. It means the recording of ethnic identities, the setting of boundaries separating "affected" groups from "unaffected" groups, arguments among the as yet "unaffected" whether they, too, do not have claims to be considered "affected." It turned out that the effort to make the Negro equal to the *other* Americans raised the question of who *are* the other Americans? How many of them can define their own group as *also* deprived? The drawing of group definitions increased the possibilities of conflicts between groups and raised the serious question, what is legitimate redress for inequality?

In 1964, we declared that no account should be taken of race, color, national origin, or religion in the spheres of voting, jobs, and education (in 1968, we added housing). Yet no sooner had we made this national assertion than we entered into an unexampled recording of the records of the color, race, and national origin of every individual in every significant sphere of his life. Having placed into law the dissenting opinion of *Plessy* v. *Ferguson* that our Constitution is color-blind, we entered into a period of color- and group-consciousness with a vengeance.

Larger and larger areas of employment came under increasingly stringent controls so that each offer of a job, each promotion, each dismissal had to be considered in the light of its effects on group ratios in employment. Inevitably, this meant the ethnic group of each individual began to affect and, in many cases, to dominate consideration of whether that individual would be hired, promoted, or dismissed. In the public school systems, questions of student and teacher assignment became increasingly dominated by considerations of each individual's ethnic group: Children and teachers of certain races and ethnic groups could be assigned to this school but not to that one. The courts and government agencies were called upon to act with ever greater vigor to assure that, in each housing development and in each community, certain proportions of residents by race would be achieved, and a new body of law and practice began to build up which would, in this field, too, require public action on the basis of an individual's race and ethnic group. In each case, it was argued, positive public action on the basis of race and ethnicity was required to overcome a previous harmful public action on the basis of race and ethnicity.

Was it true that the only way the great national effort to overcome discrimination against groups could be carried out was by recording, fixing, and acting upon the group affiliation of every person in the country? Whether this was or was not the only way, it is the way we have taken. Why we have taken this particular course and whether it is necessary is the subject of this book.

2

Affirmative Action in Employment: From Equal Opportunity to Statistical Parity

IN 1972, a committee of the House of Representatives was holding hearings on a bill to limit the busing that could be ordered by a court in cases of discrimination. There was a poignant exchange between Congressman James G. O'Hara of Michigan and Clarence Mitchell, the veteran Washington representative of the NAACP. Congressman O'Hara said:

I am deeply grieved, of course, that I find myself on opposite sides of this issue with you. . . . I remember working with you on the [open housing legislation of 1968]. . . . There was reason to believe that there were a lot of people in my district who were opposed to it, but I feel comfortable with it because I believe very strongly in the right of every person to make his own life. . . .

But I don't really agree with proposals that would assign quotas . . . in any way. I remember you and I differed on the Philadelphia Plan. That is the only other occasion I remember other than this present one.

I am prepared to take whatever legal action is necessary to make sure that the schools are open to everyone who lives in the community, and that the community is open to every person who wants to live in it. But I cannot go along with an idea that says if a community and the schools do not work out to be the right proportions, black and white, then by George, we are going to make them come out to those proportions. . . .

Mitchell answered:

I am upset, too. As you have said, my recollection includes a very wonderful assocation that we had when we met in your office night and day with others of your colleagues, working for the passage of the 1964 act, and when we labored with the 1968 act, and which I know was something of a political problem for you.[1]

A few months earlier, another liberal representative, congresswoman Edith Green of Oregon, had been explaining to Congressman Augustus Hawkins of California in the House why she could not support the proposal to increase the powers of the Equal Employment Opportunity Commission without an amendment prohibiting quotas:

Title VII of the Civil Rights Act has always prohibited the establishment of quotas. During the legislative history of the Civil Rights Act it was clearly the Congressional intent not to bring about civil rights for some by denying civil rights to others. . . .

I talked to the Chairman of the Committee . . . and . . . said it would be impossible for me to support the Committee bill . . . without . . . a Congressional prohibition against . . . any quota system.

Let me tell you of three instances . . .

In my own city of Portland, we have a ship conversion plant.

In the Portland area we have, perhaps, 5 or 6 percent black population. This ship conversion plant has records to prove they have employed 15 percent minority people. As a matter of fact they have carried on an active recruitment program—seeking out members of minority groups.

The Contract Compliance Office in San Francisco came into Portland, and they said they would not be eligible for any Federal contracts unless they would have 15 percent minority employees in every single job category. . . .

There was absolutely nothing that this ship conversion plant could do to satisfy the Office of Contract Compliance in San Francisco unless they followed their orders. This required the "dumping" of labor contracts—of negotiations which had been made; seniority rights were ignored. All this was never the intent of the Civil Rights Act, and it was never the intent of the Congress. . . .

. . . A year ago last December a group of Oregon parents who are stationed in Washington, D.C., by the Department of the Military came into my office to talk about the situation in the schools which their children attend. . . . [One of the complaints was that in three months one class had had seven substitute teachers.]

I said, "Well, how can that be?"

She said, "Under the Skelly-Wright decision we had to have a quota of black and white teachers and as a regular teacher we cannot hire a white teacher. We must hire as a regular teacher a black teacher." No qualified black teacher is available for this position. They are already teaching in other schools. . . .

A third instance: A teacher here in the District schools—whom I know very well—asked for a transfer to another high school because they had moved out close to another high school. She applied, and the principal who received her application said they could not hire her.

She said, "Be very candid with me. Is my race against me?"

And the principal said, "Yes. . . . A quota has been set up. . . ." [2]

I

Clearly, in 1972, something new was happening. It is not easy to discern just what it was amidst the welter of charge and countercharge which surrounds these emotional issues. Thus, a diligent reader of the *New York Times* in this period would have been convinced that the Nixon Administration had decided to abandon all efforts to enforce the developing pattern of law which required such actions as the setting of quotas by race for employment in cases of discrimination; the setting of goals and timetables for employment in all cases of employers with Federal contracts, whether or not they had ever discriminated; the elimination of tests which showed disproportionate passing rates between the races unless they could meet extremely high standards of validity; and the like. Such a reader would have found such stories as "Nixon Eases Drive for Minority Jobs—Ends Pressure of Federal Agencies to Spur Hiring of the Disadvantaged" (February 28, 1970); "U.S. Rights Panel Finds Breakdown in Enforcement—Hesburgh Says Nation Is on 'a Collision Course' Unless Government Gets Strict—Warns on Credibility—Commission, in Report, Calls for Leadership by Nixon in Behalf of Racial Justice" (October 13, 1970); "6 Rights Lawyers Quit Justice Unit—Score Nixon on Racial Policy and Support McGovern" (May 11, 1972); "U.S. Equal Opportunity Drive Scored in Nader-Group Study" (June 25, 1972); "Nixon Held Likely to Drop Program of Minority Jobs—Is Reported Ready to Scrap Philadelphia Plan for Construction Industry—Quotas Being Reviewed—Fletcher, Former U.S. Aide Who Administered Policy, Denounces President" (September 4, 1972)—a first-page, lead story followed up by an inner page denial a day later. In this case, the *Times* was not to be dissuaded: On September 19, it again reported: "Future of Philadelphia Plan for Minorities Is in Doubt; Critics of Nixon Fear Scuttling."

Despite this, later stories revealed, at least to the untutored eye, that there was a vigorous and ever more effective enforcement of ever more stringent requirements for nondiscrimination. So, for instance, in January 1973, AT&T, under siege from the Equal Employment Opportunity Commission, signed a settlement giving tens of millions of dollars in back pay to minority and women employees and accepting stiff goals for the upgrading and employment of these groups. A strange aspect of the case, to the untutored observer, was that AT&T did not admit to, nor did the government require it to admit to, any act of discrimination. On December 21, 1973, one could read that the once-threatened Philadelphia Plan was now being imposed by the Department of Labor on the contractors of Chicago: "Minority Hiring Plan Is Imposed on Chicago's Building Industry." On March 20, 1974, one could read that "Truckers Pledge Minority Hiring—7 Big Companies, Faced by Government Suit, Accept 'Goals' of 33⅓ to 50%"—this by the action of the Department of Justice. And from 1971 on, there was increasing concern in the academic community over the ever more insistent demand by the Department of Health, Education and Welfare for the setting of goals and timetables by universities for employment. During the entire period, there were extended struggles over the demands by Federal agencies for detailed records of applicants identified by race and ethnic group and of how they fared in the process of seeking appointment; but in every case, after efforts at resistance, the universities succumbed. They spent great sums of money in drawing up records they had never before bothered to keep in order to respond to a general presumption by government agencies of discrimination of which they had not been accused. The Federal government, responding to the concerns of Jewish groups before the 1972 election, insisted that it would not impose quotas; yet "goals" and "timetables" are today negotiated daily by its agencies, as to whether or not any showing of discrimination has been made.

The costs of this new plague of legal proceedings were enormous and were, of course, borne by the American taxpayer—on the government side through an expanded employment of lawyers, on the side of the corporations and universities, one assumes, through increased charges to customers and students. As only one example of the growth of the Federal establishment dealing with equality: In 1972, the Equal Employment Opportunity Commission had a staff of thirty lawyers; with increased powers given by Congress in 1972, the staff of lawyers rose to 220. It is now 300.[3]

How does one explain the paradox of an informed public opinion—if we take the *New York Times* to stand for that—convinced that not enough is being done to combat discrimination in employment, with the reality of ever larger budgets and staffs for enforcement agencies, ever severer penalties, ever broader remedies applied by government to overcome a presumption of job discrimination?

There is, of course, one dominant interpretation of this apparent paradox: that discrimination in American life is so deeply embedded in the minds and practices of individual Americans and their institutions that the most extended, direct, and determined remedies are necessary to root it out. Thus government agencies may, at first, hope to solve the problem by simple nondiscrimination, but they discover in the course of their work that they must go further. This is the impression one gains, for example, from writings on black and minority employment or from the articles in law journals which explain ingeniously how new remedies may be applied against ingrained discrimination or "institutional racism" (as it is so commonly called). Indeed, not only is it believed that there is an ingrained prejudice and racism in American society which is resolutely opposed to every possible advance by minority groups, it is also believed by a large body of informed opinion that the actions of government to which I have referred are illusory for the most part, a sham, and, in any case, ineffective.

It is for this reason that the Federal government and its

agencies engaged in the fight against discrimination—the Office of Federal Contract Compliance of the Department of Labor, with its branches spread through every government agency; the Department of Justice; the Equal Employment Opportunity Commission created by the Civil Rights Act of 1964—are regularly denounced for inefficiency, lack of good faith, incompetence, and refusal to carry out the law. Some of the agencies of government regularly denounce others. Thus, the EEOC criticizes agencies engaged in signing contract compliance agreements for not being strict enough. Denouncing all of them is the Civil Rights Commission, which is a more independent agency than the independent agencies (though its members, too, are appointed by the President), for it stands as a kind of Cassandra declaiming over the inadequate efforts of civil rights enforcement.

Those most active in the field of civil rights, who study the state of Federal enforcement of these rights and who make the news as to what that state is, seem convinced not only that government has done very little, but that the little that has been accomplished is always in peril of being given up, owing to the widespread resistance of racist whites in the North, liberal and illiberal alike, who seemed happy enough to impose civil rights on the South but energetically resist it when it comes to their own area. Thus, those active in the fight that led to the historic Supreme Court decision of 1954 gathered on the occasion of the twentieth anniversary of the decision. They were simultaneously dedicating a new center for civil rights, set up by the Reverend Theodore M. Hesburgh, president of Notre Dame and former long-time chairman of the United States Commission on Civil Rights, with a grant from the Ford Foundation. The *New York Times* reported:

. . . It was all tempered by the frustrating realization that the historic victory won in 1954 has also faded with time and that the civil rights war is hardly over. The anniversary of Brown

39

finds the forces of civil rights in disarray, uncertain where to move next. As they see it, the movement has struck the bedrock of opposition and public inertia.

Already some of the gains of the last two decades have been eroded by a host of complex and unforeseen "second generation" problems. . . .

"While there has been some improvement, we have not gone very far," commented John A. Buggs, staff director of the Civil Rights Commission. "If you look at the South, the form of equality has been given to blacks, but the substance is not there." Mr. Buggs, at a celebratory occasion, was willing to go farther in conceding minimal change than he normally does in his CRC reports. But he did not concede much, and his words echo the beliefs of many, ". . . but the substance is not there." [4]

The *Boston Globe,* another leading liberal newspaper, was not to be outdone in appropriate gloom: "Desegregation Fervor Flags as Battleground Shifts North," reported a headline. ". . . school desegregation in the 1970's will have to be fought in the great urban areas of the North, which give every appearance of presenting a far more hostile thicket than the supposedly 'bigoted' expanses of the South." [5]

How is it possible that such disparate views as to what is happening in our country could be current at the same time? On the one hand, a view which sees the defeat of twenty years of effort, and on the other, a view—held by Congressman O'Hara and Congresswoman Green—which sees a radical and disturbing attack on the principles of justice and equal treatment in the campaign against discrimination. (It may be pointed out that the two members of Congress lost: No effective antibusing provision has been passed by Congress, and the powers of EEOC were substantially expanded. True to form, these victories brought no joy to the advocates of busing and stronger enforcement of antidiscrimination laws.)

One place to begin our exploration of this dilemma is to ask, concretely, what has happened to the condition of minority groups between the early 1960s, when a great national effort was begun to improve this condition, and the early 1970s, when these gloomy assessments by civil rights stalwarts conflicted with equally gloomy assessments by congressmen of the bludgeoning tactics of government agencies? This, too, is a much-disputed issue, as may be seen by the debate that followed an article on minority progress in *Commentary* in 1973.[6] A similar debate occurred when it appeared a few years ago that young husband-wife Negro families in the North and West had attained rough income equality with similar white families. It was argued that this only reflected the fact that more Negro wives worked; that it was temporary; that it was in other ways deceptive.

The figures have held up, and Negro progress during this period was indeed marked. In 1969, the median income of Negro husband-wife families, with heads under 35, in the North and West, was 91 percent of the median income of white families of the same type. In 1971, it was 93 percent. If both husband and wife worked—an increasingly common development for whites, as it has been for blacks—the figures for the two years were 99 and 101 percent, respectively. Nor was the South that far behind: For all husband-wife families under 35, the Negro family income was 84 percent of the white median in 1972. It was true that total figures showed little progress in income: Thus, the total black percentage of white median, which was 54 percent in 1959 and rose to 66 percent in 1969, dropped back to 58 percent in 1972. But the depressed total figure was, in large part, the result of the rapid increase of female-headed families (from 23.7 percent of all black families in 1965 to 34.0 percent in 1974) among blacks. Among black husband-wife families, for the nation as a whole, median family income as a percentage of white did rise steadily from 62 percent in 1959 to 85 percent in 1972. There was a rapid rise of blacks and other races into occupations of greater security and higher status (the following figures are for "Negroes and other races"—Chinese and Japanese, principally—but since this group as a whole is over 90 percent

AFFIRMATIVE DISCRIMINATION

Negro, the figures do reflect changes primarily among Negroes):
Only 4.9 percent of male "Negroes and other races" were
professional and technical workers in 1963 and 8.2 percent
in 1973, a much more rapid increase than for whites (12.6
to 14.2 percent). White-collar workers rose overall from 15.3
percent of male "Negroes and other races" to 22.9 (whites:
40.6 to 41.7 percent). Craft workers rose from 10.7 to 14.9
percent (white: 20.3 to 21.5 percent).

Changes among Negro women were even more striking:[7]

TABLE 1 *

	1963		1973	
	Negro and other races	White	Negro and other races	White
White-collar workers	21.2	60.8	41.5	63.3
Professional and technical	7.8	13.5	12.0	14.9
Clerical workers	10.2	33.8	24.4	35.7
Private-household workers	34.3	4.9	12.4	2.9

Let us supplement this simple sampling of statistics with
reports from two sophisticated econometric analyses, one of
which directly tries to estimate what amount of the difference
between Negro and white income might be owing to
discrimination. Richard Freeman writes:

The income and occupational position of black workers improved significantly relative to those of whites in the sixties . . .
women, young men, young male college graduates . . . experienced especially large economic gains. By the 1970's black
women had earnings as high, or higher than, comparable white
women in the country as a whole; young black male college
graduates earned as much as their white counterparts; and the
black-white income ratio for young men in general was 0.85. . . .
As a result of increased incomes for highly educated and skilled
black workers, the historic pattern of declining black-white income ratios with ascending skill no longer prevails . . . these ad-

* See footnote 7, p. 55.

vances suggest that traditional discriminatory differences in the labor market are abating rapidly.[8]

Even more striking are the results of an econometric analysis of young black and white workers in the Sixties, which took into account the relative "endowment" of the two groups—father's occupation, highest grade in school, and achievement in school as measured by tests. Hall and Kasten write:

Holding these constant, we find that young black men entering the labor market were just as likely as whites to find high-paying jobs and just as likely to escape from bad jobs. . . . The sixties saw the nearly complete elimination of racial bias in the way that the labor market assigned individuals to occupations. . . . [9]

Of course, there were other aspects of the black position that were extremely unfavorable. There was the increase in female-headed families, the persistence of higher unemployment rates for blacks than whites, and, as Hall and Karsten emphasize, the fact that black achievement in school had not improved relative to white. But as far as the overall position of blacks and whites in the labor market was concerned, the fact was, as Freeman wrote: "While black-white differences have not disappeared, the convergence in economic position in the fifties and sixties suggests a virtual collapse in traditional discriminatory patterns in the labor market." [10] It is against this background in reality that we must explore the development of policies dealing with discrimination in the later 1960s and 1970s.

II

One place to begin is with the Civil Rights Act of 1964. In the wake of the assassination of President Kennedy and the harrowing and violent resistance in the South to the exercise of simple political rights by blacks, the nation decided, in an act of sweeping power, to finally fulfill the 100-year-old promise of the Emancipation Proclamation. The Act dealt with the right to vote (Title I), to use places of public accommodation (Title II), with the desegregation of public facilities (Title III), with the desegregation of public education (Title IV), with the expansion of the powers of the Commission on Civil Rights (Title V), with nondiscrimination in Federally assisted programs (Title VI), and, most significantly for employment discrimination, with equal employment opportunity (Title VII). The Act could only be read as instituting into law Judge Harlan's famous dissent in *Plessy* v. *Ferguson:* "Our Constitution is color-blind." Again and again, one could read the sonorous phrases: no discrimination or segregation "on the ground of race, color, religion, or national origin" (Titles II and VI), "on account of his race, color, religion, or national origin" (Title III), "by reason of race, color, religion, or national origin" (Title IV), "because of such individual's color, religion, sex, or national origin" (Title VII). Following the pattern of treatment of ethnic differences that had emerged from American experience, as described in Chapter I, the Act was understood as granting not *group* rights but *individual* rights. Two provisions, among others, were inserted in Title VII to protect individual rights:

703 (h) . . . it shall not be an unlawful employment practice . . . for an employer to give and act upon the results of any professionally developed ability test provided that such test, its administration or action upon the results is not designed, intended or used to discriminate because of race, color, religion, sex or national origin. . . .

703 (j) Nothing contained in this title shall be interpreted to require any employer . . . to grant preferential treatment to any individual or to any group because of the race, color, religion, sex, or national origin of such individual or group on account of an imbalance which may exist with respect to the total number or percentage of persons of any race, color, religion, sex, or national origin employed by any employer. . . .

And the statements made at the time of the debate on the

bill to establish the legislative history and Congressional intent seemed clear and unambiguous:

> The Civil Rights Act's floor managers in the Senate, Senator Joseph Clark of Pennsylvania and Senator Clifford Case of New Jersey, stated that ". . . It must be emphasized that discrimination is prohibited as to any individual. . . . The question in each case is whether that individual was discriminated against." [110 Cong. Rec. 7213.] Senator Clark responded to the objection that "the bill would require employers to establish quotas for non-whites" with the flat statement "Quotas are themselves discriminatory." [110 Cong. Rec. 7218.] Senator Humphrey, the majority whip, noted that "The proponents of the bill have carefully stated on numerous occasions that Title VII does not require an employer to achieve any sort of racial balance in his work force by giving preferential treatment to any individual or group." [110 Cong. Rec. 12723.] Senator Williams, explaining Sec. 703 (j), stated that it would "specifically prohibit the Attorney General, or any agency of the government, from requiring employment to be on the basis of racial or religious quotas. Under [this provision] an employer with only white employees could continue to have only the best qualified persons even if they were all white." [110 Cong. Rec. 14331.] [11]

The will of Congress and its laws must be interpreted both by administrative agencies issuing guidelines and by courts interpreting legislation and guidelines since litigation inevitably accompanies all new legislation. Thus, one may think of the law as not only the specific law as passed by Congress but as part of a troika, with the agency interpretations on the one hand and the legal interpretations on the other. The law, apparently, generally comes off a bad third in this troika, because, after all, the congressmen do not enforce the law—the agencies do —and the congressmen do not interpret the law, unless they are willing to undergo the elaborate ordeal of legislation again —the courts do.

In the case of minority employment, there was another and independent source of law: Executive orders, which go as far back as No. 8802, in 1941, issued by President Roosevelt, and ordering an end of discrimination in defense industries. Under

Presidents Truman and Eisenhower, further executive orders were issued, extending the ban on discrimination by government contractors and setting up various bodies to oversee and enforce it. Executive Order No. 10925, issued by President Kennedy, for the first time used the term "affirmative action." Contractors were now to act affirmatively to recruit workers on a nondiscriminatory basis. But the capstone of the structure is Executive Order No. 11246, issued by President Johnson in 1965. "Affirmative action" was not further defined. Presumably, it meant such things as advertising the fact, seeking out qualified applicants from sources where they might be found, and the like. Executive orders, just as laws, breed their attendant throng of regulations and guidelines, which the contractor in search of government business must attend to as carefully as (indeed, more carefully than) the executive order itself. By the time we reach guidelines, the "Executive," in the form of the President and his advisors, is far away: The permanent or semipermanent officials engaged in the program of contract compliance are the chief formulators of guidelines.

In May 1968, the Department of Labor, in which the Office of Federal Contract Compliance is housed, issued further regulations expanding upon this modest phrase, "affirmative action." A "written affirmative action compliance program" is required from every major contractor and subcontractor (more than 50 employees and a contract of $50,000 or more).

A necessary prerequisite to the development of a satisfactory affirmative action program is the identification and analysis of problem areas inherent in minority employment and an evaluation of opportunities for utilization of minority group personnel. The contractor's program shall provide in detail for specific steps to guarantee equal employment opportunity keyed to the problems and needs of members of minority groups, including, when there are deficiencies, the development of specific goals and time-tables for the prompt achievement of full and equal employment opportunity. Each contractor shall include in his affirmative action compliance program a table of job classifications. . . . The evaluation of utilization of minority group personnel shall include . . . an analysis of minority group representation in all categories. [Title 41, C.F.R., 60–1.40]

It was not at all clear at this point that a "deficiency" might be an insufficient number of workers at some level of employment of some specific group compared with the number of that group in the population. "Utilization" was used but not "underutilization," and "specific goals and timetables" were required but only for the "prompt achievement of full and equal employment *opportunity*" (my italics). The shift from "opportunity" to "representation" had not yet occurred, nor was it clearly stated that only some groups were the specific object of government concern. But there was a requirement that the contractor was to file "complete and accurate reports on Standard Form 100 (EEO–1)" (60–1.7 [a]), and that form, required by the Equal Opportunity Commission from employers covered by Title VII of the Civil Rights Act and government contractors and subcontractors covered by Executive Order No. 11246, did require a report on employees who were "Negroes," "Orientals," "American Indians," and "Spanish Americans." "Spanish Americans" were defined as those of "Latin American, Mexican, Puerto Rican, or Spanish origin." "Oriental" was not defined. (The EEO–1 form later changed from the usage "Spanish American" to "Spanish surnamed American.") On the basis of such a report, a "deficiency" in numbers employed in given categories might be found, but the May 1968 guidelines nowhere suggest that this is a breach of affirmative action requirements.

The next set of guidelines, dated February 5, 1970, requires more from government contractors in the way of affirmative action:

An affirmative action program is a set of specific and result-oriented procedures to which a contractor commits himself to apply every good faith. The objective of these procedures plus

47

such efforts is equal employment opportunity. Procedures without effort to make them work are meaningless; and effort, undirected by specific and meaningful procedures, is inadequate. [41, C.F.R., 60–2.10.]

The meaning of this new language is not completely clear, though ominous. "Opportunity," it seems, is being redefined as "result." But the specific definition of a "utilization analysis" in the guidelines, while expanded, is still oriented to "opportunity."

It is in the next set of guidelines, of December 4, 1971, that we have the creation, for purposes of Federal contract compliance, of a special category of "affected class," and a special requirement for determining whether the members of this class are "underutilized," and requirements for measures that will correct this "underutilization." The specific language involved in this order, the basis of affirmative action as required by the government from just about every substantial employer in the country, reflects the government's understanding of the causes of the differential distribution of ethnic groups in employment, and its expectations of what the pattern of employment in a nondiscriminatory society would look like.

Repeating the language of the earlier 1970 guidelines, that ". . . procedures without effort to make them work are meaningless; and effort, undirected by adequate and specific procedures, is inadequate, . . ." it now continues:

An acceptable affirmative action program must include an analysis of areas within which the contractor is deficient in the utilization of minority groups and women, and further, goals and timetables to which the contractor's good faith efforts must be directed to correct the deficiencies and, thus to increase materially the utilization of minorities and women, at all levels and in all segments of his work force where deficiencies exist.

The point of this pronouncement is that equal employment opportunity must now be redefined, against its plain meaning, not as opportunity, but result. "Procedure" and "effort" alike are inadequate without "result." The employer is required to undertake

> . . . an analysis of all major classifications at the facility, with explanations if minorities or women are currently being underutilized in any one or more job classifications. . . . "Underutilization" is defined as having fewer minorities or women in a particular job classification than would reasonably be expected by their availability. . . . [41, C.F.R., 60–2.11.]

Some guidance is given on how to determine "underutilization" (60–2.11): In effect, the census is now to determine what is discrimination and what is affirmative action. That all this is still called "equal employment opportunity" is simply another example of the misnaming of reality in an age in which words are easily distorted into their opposites. This is the last in the series of guidelines under which the Office of Federal Contract Compliance and its multifarious branches in every agency of the executive establishment operate, and the law of the land.

Legally, the Equal Employment Opportunity Commission operates under greater restrictions than the Office of Federal Contract Compliance: After all, there is still the specific ban in the law which created it against the use of any statistics of "imbalance" to require preferential treatment. However, with the assistance of the courts, simple imbalance—under which the EEOC can require nothing—can be redefined as itself a showing of discrimination, which permits the EEOC or the Department of Justice to require everything: back pay for classes of individuals who themselves have suffered no discrimination, the setting of quotas for employment of individuals of specific groups for given jobs, and the like. The question one may ask is: When does imbalance, under which one can do nothing, become discrimination, on the basis of which one can do everything?

The development of the law here, as interpreted by the courts, has been deeply affected by the dogged resistance by

Southern states and localities to the granting of political rights to blacks and to the dismantling of the dual school system. The South was endlessly ingenious in devising regulations for registration that were on their face fair but that were used in many jurisdictions to deny the blacks the right to vote. In the Voting Rights Act of 1965, all subterfuges were finally thrust aside under a simple statistical rule: If 50 percent of the persons of voting age, according to the census, had not voted, radical and drastic Federal intervention and correction could now be undertaken. It was under this drastic statistical rule that the barriers to Negro voting in the South finally fell, and the political situation of the Negro was transformed. Similarly, in 1968, the Supreme Court, tired of endless delay in desegregating dual school systems, accepted a statistical rule for desegregation of schools, under which "freedom of choice" was outlawed. If too few blacks and whites took advantage of freedom of choice to shift their schools, that demonstrated unconstitutional discrimination had not been overcome, period.

Thus it was relatively easy for a statistical rule to develop in discrimination cases brought under the Civil Rights Act of 1964, for here, too, endless subterfuge was possible. Indeed, it was owing to the resistance of crafts unions to the entry of blacks that the first affirmative action programs requiring fixed statistical quotas for employment were instituted by the Federal government through its power as a Federal contractor, in the Philadelphia Plan in 1969, and many similar plans which followed. The argument could be made in the courts, and was, that either the statistical disproportion in employment was such that it served as a prima facie case of discrimination—under which the judge could impose whatever remedy seemed suitable to him, including fixed quotas for employment—or that the executive order's affirmative action provisions were not bound by the specific provision against preferential treatment on account of imbalance in the Civil Rights Act of 1964. Thus the statistical approach to proving discrimination spread from voting and the schools to jobs, even though one can, on the face of it, discern a very important distinction: Everyone, with minor exceptions, is expected to have the right to vote and is required to go to school, but jobs are based on qualifications and it is well-known that qualifications (such as education) will vary with race and ethnicity.

It was to protect the right of the employer to get qualified employees that the provision we have quoted from the Civil Rights Act of 1964, permitting the use of "professionally developed ability tests," was introduced into the law. The establishment of a statistical rule for discrimination has found as its chief obstacle this issue of how employers are to determine "qualifications," but the EEOC and the Department of Justice and the Office of Federal Contract Compliance have been rather effective in pushing aside this obstacle through other guidelines on "Employee Selection Procedures" (Federal Register, August 1, 1970).

Any test that distinguishes between individuals (which is, after all, their purpose) will also, willy-nilly, distinguish between groups. If it tests for vocabulary, or knowledge of rules, or ability to understand instructions, it will clearly be affected by the differing degrees of education and educational achievement that are characteristic of different groups at a moment in time. If it tests for nonverbal capacities, it will also, owing to the complex and subtle effects of history and culture, distinguish between groups. Even if it tests for height—as is the case for some occupational tests—it will distinguish between groups. The guidelines are to tell employers when a test that shows differences in achievement (or existence) between groups will be acceptable.

The position taken in the guidelines is that any test ". . . which adversely affects hiring, promotion, transfer or any other employment or membership opportunity of classes protected by Title VII constitutes discrimination unless: (a) the test has been validated and evidences a high degree of utility as hereinafter described; and (b) the person giving or acting upon the results of the particular test can demonstrate that alternative suitable hiring, transfer or promotion procedures are unavailable for his use."

The Supreme Court upheld these guidelines in the landmark case of *Griggs* v. *Duke Power Company* in 1971: Any test which blacks and whites passed at different rates would have to be carefully scrutinized and validated. Otherwise, it would be considered prima facie evidence of discrimination. But now, what is "validation" according to the guidelines that the Supreme Court said "is entitled to great deference"? Analysis of the detailed EEOC requirements for validation by various authorities has led to the conclusion that just about no test that shows differential achievement really can be validated: The requirements are simply too stringent. The *Harvard Law Review,* analyzing the testing guidelines, asserts,

. . . if applied literally they would raise the cost of testing for many employers beyond tolerable limits, forcing the abandonment of testing programs which, although they may be valid, cannot be validated at any cost. . . . It is possible to read the Guidelines so strictly as to make testing virtually impossible. . . .

The Guidelines, if applied as strictly as their language allows, would encourage many employers to use a quota system of hiring. Because of the impracticality of the validation and alternative showing requirements, and the fact that the Commission will scrutinize closely "higher rejection rates for minority candidates than non-minority candidates," the easiest way for an employer to stay out of trouble and avoid exhaustive validation techniques is hire an acceptable proportion of blacks by applying a lower cutoff score to black applicants. . . .[12]

Another careful survey of the state of the law concludes:

Despite validation and evidence of utility, the presence of disproportionately lower scores for minorities may be sufficient in and of itself to find a test discriminatory. [Here there is a reference to an EEOC decision.] Another federal standard provides that the relationship between test performance and at least one criterion of job performance must be so statistically significant that only one out of every 20 sufficiently high scores occurs by chance. The standard is, however, unattainable for any known ability test. . . .[13]

The Equal Employment Opportunity Commission, supported by the Department of Justice and the Federal courts, has carefully constrained the use of tests to the point where, in fact, hardly any test may be used without legal challenge. When the testing authorities of the Equal Employment Opportunity Commission are asked for an example of a test that will pass muster as legitimate, if blacks pass at a lower rate, the only example they seem to come up with is a typing test. Even that, we should point out, may not be used indiscriminately. One cannot, for example, require a higher level of typing than is current in a group of typists already employed. Indeed, one would not be allowed to require a higher level of typing than that possessed by the poorest typist one has hired, if the higher level served to select one ethnic group or race more or less than another. Under this rule, one could guarantee that the level of typing would steadily decline.

The development of the law in these areas has become truly arcane as one Federal agency plunges ahead of another, and the other races to catch up, or the courts go beyond either. The effect is to steadily constrain any effort to set a higher standard of employment or, indeed, any standard, if it serves to have disproportionate impact, even a minor disproportionate impact, on the employment of some specific group. Any disproportion may trigger investigation and the application of stringent rules.

In effect, the EEOC is engaged in breaking the law under which it operates. "In 1970, a member of the EEOC staff told the *Harvard Law Review* that 'The anti-preferential provisions [of Title VII] are a big zero, a nothing, a nullity. They don't mean anything at all to us.' " [14] Anyone studying their record hardly needs this direct confirmation.

The testing guidelines whose impact we have summarized

are, even so, judged inadequate by the Equal Employment Coordinating Council (which brings together the EEOC, OFCC, the Department of Justice, the Civil Rights Commission, and the Civil Service Commission), and they have been trying to implement even more severe guidelines. The Civil Service Commission understandably is most concerned with the protection of the merit system. The Equal Employment Opportunity Commission, together with the Civil Rights Commission, has been engaged for some time in an attack on the Civil Service Commission's Federal Service Entrance Examination, asserting it is discriminatory because more blacks fail it than whites. The Civil Rights Commission is particularly annoyed at the Civil Service Commission because the latter, operating on the quaintly antique view that the merits of an individual are more significant than that individual's race or color in determining eligibility for a job, refuses to record the race of those taking the test. The Civil Rights Commission complained in 1973:

. . . In persisting in using the Federal Service Entrance Examination to measure the ability of approximately 100,000 job applicants for more than 100 Federal job classifications, the CSC falls short of exercising its responsibility. That the FSEE has not been properly validated to ensure that it does not discriminate against minorities is a matter that has been raised by civil rights groups and certain Federal agencies, including the EEOC and this Commission. The Civil Service Commission maintains that the FSEE is fair and non-discriminatory and that it is a relatively accurate indicator of how a person will perform on the job. . . .

What is important is whether the FSEE screens out qualified minority applicants. Since CSC does not keep records of the racial or ethnic identity of persons taking the FSEE, there is no way of knowing whether this occurs.[15]

The CSC has gone quite far in meeting the complaints of the CRC. Thus, as long ago as 1970, the CSC, according to a CRC review, had ". . . largely . . . eliminated . . . arithmetic and algebraic components of the exam." [16] In addition, as is pointed out in the CRC's 1973 report, the CSC commissioned the Educational Testing Service to undertake a 6-year study to validate parts of the FSEE. This study was a monumental undertaking: How *does* one show the "job-relatedness" of an entry level test for something like 100 different job classifications? It must inevitably be something on the order of an intelligence test. Nevertheless, this task was undertaken at what must have been enormous expense. It demonstrated that those who did better at the tests did indeed do better at their jobs, as best as one could figure out what doing better at one's job meant. It also showed that members of minority groups who did well on tests did better on jobs, just as those of majority groups did. The CRC remains unconvinced. It refers in its 1973 report to this monumental study as only "allegedly" demonstrating that ". . . people who do well on tests do equally well on the job. . . ." [17]

The disdain of the CRC for the tests developed over decades by the CSC is total. Interviewed on the proposed new testing guidelines, John A. Buggs, staff director of the CRC, said in an interview with the *National Journal,* "I don't believe there is such a thing as a merit system. Whether we can develop one is another matter, but I don't believe we have ever had one." [18] Presumably, Mr. Buggs will not be satisfied until requirements for literacy are removed from Federal tests, as requirements for numeracy have been.

The new proposed guidelines for testing emerged out of eighteen months of negotiations between the five Federal agencies on the Equal Employment Opportunity Coordinating Council and tell us what these agencies consider a nondiscriminatory means of testing for employment.

Thus to quote from the *National Journal* analysis:

One point on which [the agencies involved] all seem to agree is the virtual impossibility of validating unscored interviews.

What this means is that if any unscored interviewing procedure statistically seems to have an adverse impact on any group, employers simply will be unable to prove the interview's validity. . . .

The problem of interviewing . . . becomes particularly acute in high-level professional jobs where factors such as creativity, initiative, drive and even intraoffice personal compatibility become decisively important. . . .

If there is any showing of adverse impact, the guidelines clearly require employers to adopt quantified over unquantified hiring tests. If the employer is not happy with quantified procedures, his only alternative is to remove the adverse impact by hiring members of each group in proportion to the numbers in which they apply. . . .

. . . All requirements of educational certification or licensing used in our credentials-conscious society will have to be validated, under the proposed guidelines, if they can be shown to have the adverse impact they almost invariably do have.

Some degree requirements will make it, others almost certainly will not. Requirements for M.D. degrees probably can withstand legal challenge. . . . At the other extreme, requiring a B.A. for salesmen on the assumption that the B.A. is the sign of an articulate person is almost certainly impermissible under the guidelines. In the middle are requirements such as teaching degrees, licenses or certificates, or the requirement of a Ph.D. for many college and university teaching positions.[19]

The practice of using biographical histories also comes under attack in the guidelines. The EEOC has already ruled that to take into account a criminal record in hiring is discrimination on account of race:

Since (1), a substantially disproportionate percentage of persons convicted of "serious crimes" are minority group persons and (2), clearly it is arbitrary and therefore unnecessary to treat all "serious" convictions as being equally predictive of future employment without reference to the particular factors of a particular case, such as job-relatedness of the conviction and the employee's immediate past employment history, we hold that respondent's policy of automatic discharge for any "serious" crime discriminates against Negroes as a class because of their race within the meaning of the Act. . . . (No. 72–1497)[20]

The guidelines take this position, too, and carry it somewhat further. Trying to find an example of a criminal record that may be disqualifying, they come up with the following: ". . . a recent conviction or history of embezzlement may disqualify an applicant for a position of trust requiring the handling of money or accounts." Note the "may." [21]

Already well established by the EEOC is the requirement for a test of "differential validation." The point here is that if members of a minority group do worse on any test, and the employer still wishes to use the test, he must, at his own expense, carry through a validation study which demonstrates that the relationship between the test and performance on the job is not different for minorities from what it is for others. "Validation alone [leaving aside differential validation] may cost between $40,000 and $50,000 under favorable circumstances, according to Robert M. Guion, past president . . . of the Industrial and Organizational Psychology Division of the American Psychological Association. . . ." "Differential validation" would cost much more.

It turns out that the only expert in the field who seems to have a strong belief in this phenomenon is the director of research of the EEOC. Neither the experts of the Civil Service Commission nor independent scholars believe either that it exists or that it turns up often enough to worry about. It seems strange that elaborate validation studies should be demanded on the basis of the existence of a doubtful phenomenon. But this is not the full extent of the fantasy world constructed by the antidiscrimination agencies. Suppose an employer, using a test, does not have enough minority employees to test for this doubtful phenomenon of differential validity? Then, according to the EEOC research director,

. . . the employer . . . will be expected to hire enough people in the underutilized, adversely affected group to do the differential prediction study. . . .

Furthermore, [the research director] said, "the employer may not perpetuate the exclusion of minorities by . . . saying that he is choosing from the available pool of 'qualified' candidates. That simply would beg the question of how do you test for qualifications."

In other words, given adverse impact, there is a presumption of discrimination that the employer must bear the burden of disproving.[22]

III

The purpose of this determined and dogged resistance to any form of employment selection which leads to any but proportionate representation in a work force is simply that—to ensure proportionate representation of minority groups in a work force. "Affirmative action" originally meant that one should not only not discriminate, but inform people one did not discriminate; not only treat those who applied for jobs without discrimination, but seek out those who might not apply. This is what it apparently meant when first used in executive orders. In the Civil Rights Act of 1964, it was used to mean something else—the remedies a court could impose when some employer was found guilty of discrimination, and they could be severe. The new concept of "affirmative action" that has since emerged and has been enforced with ever greater vigor combines both elements: It assumes that everyone is guilty of discrimination; it then imposes on every employer the remedies which in the Civil Rights Act of 1964 could only be imposed on those guilty of discrimination.

Affirmative action has developed a wonderful Catch-22 type of existence. The employer is required by the OFCC to state numerical goals and dates when he will reach them. There is no presumption of discrimination. However, if he does not reach these goals, the question will come up as to whether he has made a "good faith" effort to reach them. The test of a good faith effort has not been spelled out. From the employer's point of view, the simplest way of behaving to avoid the severe penalties of loss of contracts or heavy costs in back pay (to persons selected at random who have not been discriminated against, to boot), such as have already been imposed on AT&T and other employers, is simply to meet the goals.

How the EEOC is likely to look upon a failure to meet the "goals" that have been set (and that the employer must set) is grimly set forth in the *7th Annual Report:*

As the Commission stated in Decision No. 72–0265, Title VII imposes an affirmative duty on employers and unions to end the chilling effects of past discrimination and that a continuing lack of Negro applicants for once all-white jobs only indicated that, "The effectiveness, thoroughness, and frequency of whatever efforts the respondent was making to inform Negroes that it no longer discriminates against them fall short of what is necessary." [23]

There is a simple solution to Catch-22: proportional hiring, quotas; and every employer worth his salt knows that is the solution that the EEOC and the OFCC and the rest of the agencies are urging upon him, while they simultaneously explain they have nothing of the sort in mind.[24]

It is hardly worth trying to catch the agencies in this transparent ploy: The evidence is overwhelming, as "affirmative action" plans are turned back under threat of canceled contracts and legal procedures. Interestingly enough, the fullest evidence is from the academic community, and one wonders why. Business is under equal or more pressure and also has had to pay heavy sums in settlement of back pay claims. But it has fought less. Is this, perhaps, because business has a greater sense of guilt because its actions have indeed been discriminatory? Because it is diffident over the fact that its desire to select the most efficient and competent employees will serve only to supply better telephone services or make better cars? It is hard to see why institutions of higher education should in particular have the freedom to seek out the best, regardless of whether business does. John Gardner believes that every pursuit has its own excellence and every excellence should be encouraged, and I agree. Perhaps the university has publicized its problems more because university faculty, if not administrators, are not, as yet, acquainted, as business is, with the enormously intrusive reach of government into every aspect of the workings of institutions, and are thus more outraged when they experience it. Perhaps they simply take the principle of merit more seriously. In any case, the evidence from the universities of an illicit governmental insistence that quotas be set is overwhelming.

Thus Sidney Hook quotes a letter to the president of the University of Arizona on March 31, 1971, from an HEW official who told him: "Department Heads should be advised that, in addition to the active recruitment of females, affirmative action requires that Government contractors consider other factors than mere technical qualifications."

An announcement from San Francisco State College of October 8, 1971, of an "affirmative action plan" approved by HEW calls for:

an employee balance which in ethnic and male/female groups, approximates that of the general population of the Bay Area from which we recruit. What this means is that we have shifted from the idea of equal opportunity in employment to a deliberate effort to seek out qualified and qualifiable people among ethnic minority groups and women to fill all jobs in our area.

A letter in the *New York Times* of January 6, 1972, from some Cornell University faculty members reports:

. . . that policy, as described in a letter from the President of the university to the deans and department chairmen, is "the hiring of additional minority persons and females" even if "in many instances, it may be necessary to hire unqualified or marginally qualified persons."

60

This is what appears in written communications; worse is reported from the face-to-face meetings:

At one Ivy League university, representatives of the Regional HEW demanded an explanation of why there were no women or minority students in the Graduate Department of Religious Studies. They were told that a reading knowledge of Hebrew and Greek was presupposed. Whereupon the representatives of HEW advised orally: "Then end those old fashioned programs that require irrelevant languages. And start up programs on relevant things which minority group students can study without learning languages." [25]

The result of this pressure was, of course, a flood of advertisements and recruiting letters indicating, in one way or another, "minorities or women only," and internal memoranda to the same effect. Thus, a January 12, 1973, memorandum from the Chairman of the English Department of College I of the University of Massachusetts, Boston, on faculty recruitment, reads, "At present we are authorized, in accordance with the University's strong commitment to Affirmative Action recruitment, to interview only candidates from ethnic minorities. . . ." As against the assumption of some students of minority employment, still focused on earlier data that members of minority groups are employed at lesser salaries, it has become common knowledge that minority faculty of any given level of achievement (though generally not women—in that case, the labor supply has been quite elastic) would have to be paid more, if they were to be recruited.

There is also statistical evidence of the extent to which preferential hiring to meet goals and timetables has spread through the academic community.

A survey of 162 chairmen of sociology departments revealed first, in response to the question: "Do you believe that to follow HEW Guidelines concerning the hiring of women and minorities would force you to act contrary to the Civil Rights

Act of 1964?," 17 percent said yes, 44 percent no, and 39 percent do not know. Of course, these are not lawyers, who have written both the law and the guidelines and presumably can reconcile them. Their action is perhaps more interesting than their opinions. Of these chairmen, 139 had been in a position to hire faculty members within the last two years. Of these, forty-four–32 percent–". . . reported they had felt coerced to hire a woman or a minority member regardless of whether or not he or she was the best candidate for the job." Thirty-four percent of those who felt coerced reported the coercion came from the administration, and we must assume *they* felt coerced by HEW.[26]

The first great success in the drive to impose quotas on business in the absence of findings of discrimination and simply on the basis of a requirement of proportionality came in the elaborate case mounted by the EEOC against AT&T. This case began before the agency had the legal enforcement powers given in the 1972 amendments, but it was able to undertake it because AT&T had requested the Federal Communications Commission to grant it a rate increase, and EEOC was admitted as a party, arguing discrimination by AT&T. The summary of the EEOC case, which ran to many thousands of pages, reveals a simple-minded commitment on the part of this government agency to one principle testing for discrimination: equal representation.

The summary asserts, ". . . it is absolutely clear that blacks are not randomly distributed in all jobs, . . ." as if there was any expectation that blacks or any other category should be "randomly distributed in all jobs." [27]

Speaking of entry-level jobs, it asserts:

> . . . A substantial underrepresentation of women or minorities in certain job categories manifestly cannot be attributed to their lack of skill. Absent discrimination, one would expect a nearly random distribution of women and minorities in all jobs.[28]

Absent discrimination, of course, one would expect nothing of the sort. Economists, labor market analysts, and sociologists have devoted endless energy to trying to determine the various elements that contribute to the distribution of jobs of minority groups. Some of the relevant factors are: level of education, quality of education, type of education, location by region, by city, by part of metropolitan area, character of labor market at time of entry into the region or city, and many others. These are factors one can in part quantify. Others—such as taste or, if you will, culture—are much more difficult to quantify. Discrimination is equally difficult to quantify. To reduce all differences in labor force distribution (even for entry-level jobs) to *discrimination* is an incredible simplification. It is a simplification in which the Federal agencies persist.

One striking example may be taken from the highly publicized 1970 report of the Civil Rights Commission on the Federal civil rights enforcement effort. The CRC report points out, as is well-known, that Negro employment is concentrated overwhelmingly in the lower grades of Federal employment. There had been a rapid increase in the five years between 1962 and 1967; thus, the percentage of blacks at higher levels had increased two and a half times in the Civil Service, six and a half times in the Wage Board group of employees, six times in the Postal Field Service. The response of the CRC report was ". . . the 1967 picture still reflected gross under-representation of Negroes in better paying jobs." [29]

But let us look at the matter more closely. One way of increasing the number of Negroes in the Federal Civil Service would be to recruit more actively in predominantly all-Negro colleges in the South. The report informs us that ". . . a visit by one or more federal officials is made for every 20 black students; the ratio for whites is estimated at 1:225." Certainly the Federal government has not been deficient in the *scale* of its recruitment efforts! Is there room for improvement? Another interesting figure in the report tells us that of 1400 June 1967 graduates from fifty-one black colleges, 656 had accepted jobs with government, primarily Federal.

The report also points out:

An analysis of occupational categories comprising most G.S. 15 to 18 executive positions . . . [reveals] . . . medicine and engineering—occupations long virtually closed to minority group members—make up nearly one-third of all positions. More than 50 per cent of federal executive level employees hold master's degrees or better. Again, the premium placed on higher educational attainment works to the disadvantage of minority group members, who have been systematically deprived of equal educational opportunities for generations. Other characteristics of G.S. 15 to 18 executives—long years of federal service [two-thirds of the group have more than 20 years of federal service] and age level [more than 80 per cent are 45 or older]—also shed light on the grossly inadequate minority group representation within the upper grades.[30]

If one-third of the very highest posts in the Federal Civil Service consists of engineers and doctors, and if Negroes make up less than 2 percent of all engineers and doctors, clearly, it is no simple matter to raise the proportion of Negroes in the higher Federal service. Indeed, to make up for the fact that there are very few Negro doctors and engineers (and Negro doctors at least would very often find private practice, or research or other opportunities, more attractive), the proportion of Negroes in the other two-thirds of the higher Federal service would have to go considerably above 11 percent to make up the deficiency.[31]

It is on the basis of such standards of proof that quotas are now set on businesses and on local municipalities (which were brought under the jurisdiction of the Equal Employment Opportunity Commission by the Civil Rights Act amendments of 1972) by the Federal agencies and by Federal courts. If anyone operates under the illusion that discrimination as ordinarily understood has anything to do with the setting of such quotas, examination of specific cases will disabuse him.

For example, in 1974, a Federal District judge ruled that Boston must henceforth hire one black for every white teacher until blacks formed 20 percent of the teaching force. What was the finding of discrimination on the basis of which this ruling was made? The decision pointed out that, in 1972–1973, 5.4 percent of the teachers were black, 3.5 percent of supervisors and other administrative and staff positions, while the black population of Boston was approximately 16 percent, black enrollment in the schools 33 percent. (The judge did not report— or perhaps no one told him, or he did not consider it relevant— that only 5.2 percent of the college graduates in Boston, and 1.4 percent of the college graduates of the Boston metropolitan area, were black.) The Boston school system uses the National Teachers Examination, prepared by the Educational Testing Service. It also interviews teacher applicants. In practice, however, the interview does not matter, and appointments depend on performance in the NTE. The ETS has suggested that NTE results should not be the determining criterion for who is hired. (In the nature of the case, it would be almost impossible to "validate" a test to select teachers, since so many different elements go into teaching.) ETS studies show that blacks do worse than whites on the NTE. Since a test that is not sufficiently precise in predicting teaching ability is used, since it is not in effect supplemented by interviews, and since the proportion of Negro teachers in the system is less than that of Negroes in the population, Judge Garrity ruled that there was racial discrimination in hiring teachers and that the Boston school system must hire one black for one white. This decision was made in the light of evidence that the Boston school system has a high official who works on black teacher recruitment and has made trips to colleges and conventions in an effort to recruit black teachers. The budget for minority recruiting was $60,000 in 1970–1971, and $34,000 for 1971–1972. The judge considered this effort inadequate. (*Morgan* v. *Hennigan,* 379 F. Supp. 463–5 [1974].) Of such matter are findings of "discrimination" now made and the setting of rigid quotas justified.

I V

It may then be granted that the Federal civil rights enforcement agencies, with their scheme of "affirmative action" based on an estimate of "underutilization," and the courts, with their strange definitions of "discrimination," are engaged in a process of requiring all the major employing institutions in the country to employ minorities in rough proportion to their presence in the population. While this may be of concern to the legal specialist or the moralist, how can it be anything but a matter to applaud for the social analyst concerned for the good of the country, for the welfare of the groups within it, and for the future of the relations of the groups that make it up? We are acquainted with the despair of the ghetto, a despair that broke out in riots in the 1960s, in extremist political demands accompanied by terrorism at the turn of the 1970s, and in an enormous increase in crime which has made life in the cities miserable and dangerous. Whatever the means, can it be denied that a process of requiring equal representation of employment is good not only for the minority groups but for the country?

Let me first clear away one issue that is generally raised in the discussion of affirmative action: I oppose discrimination; I fully support the law; I think there is nothing which so degrades a nation as the exclusion of one group from its opportunities—including employment and education and political participation and access to the multifarious benefits of government—because of race, color, national origin, religion, and, I will add, as the law does, sex, although that is not at issue here. But what I have been describing and criticizing is not an attack on discrimination. Where the EEOC takes up a case of discrimination and gets a job and compensation for the victim, I applaud it. The fact is that much of the work of the government agencies has nothing to do with discrimination. One may review these enormous governmental reports and legal cases at length and find scarcely a single reference to any act of discrimination against an individual.

The downgrading of acts of *discrimination* in the legal and administrative efforts to achieve equality for blacks and other minority groups in favor of statistical pattern-seeking has some important consequences. It is one thing to read that an up-standing, hard-working, and ambitious young man has been turned down for a job, or a school admission, or a house because he is black. It is quite another to read that the percentage getting such and such a job, or buying houses in this place, or being admitted to this program is thus and so. In the latter case, one does not know why the percentage is the way it is—nor is there any necessary reason given why it should be higher or lower. The sense of concrete evil done which can, and does, arouse people disappears. If any concrete connection with the Negro condition is made, it cannot be that this and that qualified black has been denied a job or admission—for the evidence is now all too clear that the qualified (and a good number of the unqualified, too) get jobs and admissions. The concrete connection with an individual's personal fate that is likely to be made tends to be in this form: If more blacks were given these jobs, perhaps less would be on the street, or drug addicts, or killing unoffending shopkeepers. It is one thing to be asked to fight discrimination against the competent, hard-working, and law-abiding; it is quite another to be asked to fight discrimination against the less competent or incompetent and criminally inclined. The statistical emphasis leads to the latter. Undoubtedly even those of lesser competence and criminal inclination must be incorporated into society, but one wonders whether this burden should be placed on laws against discrimination on account of race, color, religion, or national origin.

The emphasis on statistics, rather than personal discrimination, raises another problem. The argument from statistics *without cases* is made not only because it is an easier argument to make, and will lead to more sweeping remedies, but also for two other reasons:

First, those who make it believe that there is such a deeply ingrained prejudice in whites, leading to discrimination against blacks and other minorities, that it can be assumed prejudice is the operative cause in any case of differential treatment, rather than concern about qualifications. To this assumption there can be no answer. One can only, as an individual, search one's own motives and actions, and those of the institutions and bodies with which one is involved. The ordinary public opinion surveys do show a substantial decline in prejudice, but certainly no one who believes in the persistence of an ingrained and deeply based prejudice which will make itself felt in every situation will accept such admittedly crude and broad-brush evidence. There is no answer to the argument that every case of differential treatment must be based on prejudice since we are all prejudiced: We will each make our own response to this, depending on our experience and on our sense of the motives of ourselves and our fellowmen. Clearly, the decisions of the anti-discrimination agencies are the least charitable possible. But they have support in liberal opinion, which continually insists on the guilt for racist actions of each and every one of us.

However, there is another justification for the statistical approach which may be easier to deal with in open discussion: This is the justification from "institutional" causes. A reference to "institutional forms of exclusion and discrimination" occurs in an official document of the Office for Civil Rights of HEW.[32] We also know the term as "institutional racism." This term has not been subjected to the analysis it deserves. It is obviously something devised in the absence of clear evidence of discrimination and prejudice. It suggests that, without intent, a group may be victimized. Racism, in common understanding, means an attitude of superiority, disdain, or prejudice toward another person because he is of another race, and a philosophy or ideology that justifies such attitudes on the basis of the inferiority—genetic, cultural, moral, or intellectual—of a race. The rise of the popularity of the term "institutional racism" points to one happy development, namely, that racism pure and simple is less often found or expressed. But the rise of this term has less happy consequences in that it tends to assume that all cases of differential representation in an institution demonstrate "institutional racism." But each institutional form of exclusion must be judged in its own terms. Thus whether the requirement of a high school diploma, or a grade in a test, or employment on the basis of some skills or talents is "institutional racism" and to be corrected must depend on the judgment of whether it is justified in terms of the end of the institution; whether, even if not fully justified, it is convenient (in which case, a sound public policy might call for its change, but without the implication that it had any racist intent); or whether it is actually designed to exclude some minority group, in which case it is clearly unjust and illegal. The term "institutional racism," however, tends to push the decision as to what we deal with increasingly toward the intepretation, as we have seen, that *any* institutional effort to make distinctions is unjust or illegal.

Against the argument that, whatever the moral or legal faults or these procedures, they are socially good, I would make three points:

First, they became institutionalized and strengthened at a time when very substantial progress had been made, and was being made, in the upgrading of black employment and income, a progress that had, oddly enough, taken place without benefit of such extreme measures (see pp. 41–43). Second, it is questionable whether they reach in any significant way the remaining and indeed most severe problems involved in the black condition. And third—which returns us to the theme of our first chapter—unnecessarily and without sufficient justification, they threaten a desirable, emergent pattern of dealing with ethnic differences we have described in Chapter I, one which does not give them formal, legal acknowledgment, neither encourages nor discourages group allegiance and identification, and treats every individual as an individual and not as a member of a group. This is now threatened by public policies which emphasize rigid lines of division between ethnic groups and make the ethnic characteristics of individuals, because of public determinations, primary for their personal fate.

I have already given some evidence on the marked progress in the economic position of blacks which was evident before the most extreme and still current form of affirmative action was established in 1971. A substantial measure of income equality has been reached in North and West for younger, complete families, and we are close to it in the South.

In addition to this change in the income position of blacks, there were remarkable changes in occupational distribution. The causes of these changes were numerous: the improved economic situation, the improved education of blacks giving them access to more jobs,[33] the population movement from South to North and from rural and small town to large city, the pressure of the black revolution and the changing attitudes of whites, and the actions of businesses and colleges and universities in increasing opportunities for minorities. The legal measures of proportional representation we have described could not very well have had much effect before 1971. They seemed to have been instituted just at the point when black progress, economically and educationally, seemed solidly institutionalized, and was maintaining itself despite the decline of riots.[34]

My second argument against proportional representation is that I do not see how it is effective in reaching the really severe problems of the black population. Among the most serious economic problems of the black population in the past decade have been high rates of unemployment—even higher rates among the youth—and a substantial decline in participation in the labor force. Now these developments occurred during a decade of economic expansion when, in most metropolitan areas, substantial numbers of unskilled jobs were available. Consider one analysis of developments in New York City, where there has been a substantial decline in the number of blacks in the labor force:

. . . [N]either rates of unemployment by sex nor by race show the New York labor market to have been weak in early 1970, either in the New York region or in New York City. . . . [W]ages of unskilled workers are relatively high and have been rising sharply in the New York region, suggesting that there is a strong demand for workers at the unskilled level. [Yet] the Bureau of Labor Statistics reported a dramatic decline in the labor-force participation in New York among "Negroes and other races" 20 years and older from 61 percent of the population in 1970 to less than 55 percent in 1972. . . .

Any attempt to expand New York employment will have to deal with the supply as well as the demand for labor. And any attempt to deal with the supply of labor should take into consideration the possibility that at least a portion of the decline in labor-force participation is due to the existence of attractive alternatives to working.[35]

Many of the features of this analysis could be duplicated in other metropolitan centers. Since 1948 there has been a long-range decline in the labor-force participation of non-white males from 97 percent to 92 percent among 35- to 44-year-olds, from 95 percent to 87 percent among 45- to 54-year-olds.[36]

Poverty among blacks is increasingly concentrated among female-headed families. Of Negro families in poverty, those families with male heads dropped from 1,300,000 to 550,000 from 1959 to 1973; with female heads, they increased in this period from 550,000 to 970,000. As a result, in 1973, of black families in poverty, female-headed families formed almost two-thirds. It is not easy to see how this group could effectively be reached with antidiscrimination programs, in view of the levels of education among them, the nature of the jobs for which they are best qualified, and the alternative attractions of welfare.[37]

During the 1960s, when discrimination was declining, the income of blacks as a proportion of white income was rising, and the percentage of blacks in white-collar and stable blue-collar work was also rising, there was simultaneously a great increase in female-headed families among blacks, of youth unemployment, and of crime among blacks. No one has given a very convincing explanation of this tangle of pathology in the ghetto, but it is hard to believe it is anything as simple as lack of jobs or discrimination in available jobs.

At the moment of writing, we are in a deep recession, and black unemployment, as white, has gone very high. But clearly it was not the absence of jobs in the late 1960's that explained the situation of a decline in labor force participation and the existence of many jobs for which there were no takers. For example, an analysis by two authorities of the problems of black youth points out:

> Because illicit activities can be an attractive alternative to work and because few youths have family responsibilities, many are not interested in taking jobs that pay low wages and have little status. When unemployment rates were at a low in 1969 and jobs were relatively plentiful, it proved to be difficult in many cities to recruit eighteen- and nineteen-year-old youths for public employment at minimum wages, even though unemployment for males of this age continued to be high. . . .
>
> The limited evidence does not disprove the notion held by many employers that youths are reluctant to work at the going wage. . . .

The authors do go on to say: "Nevertheless, the evidence suggests that a majority of teenage males have realistic wage expectations." These, of course, may be the ones who are working.[38]

It seems clear that the main impact of preferential hiring is on the better qualified—the professional and technical, who are already the beneficiaries of an income bonus on the basis of their relative scarcity; the skilled worker already employed, upgraded through governmental pressures; the unskilled worker already regularly employed, also given opportunities of upgrading. Undoubtedly some of the benefit reaches down, but the lion's share must inevitably go to the better qualified portion of the black population.

Nothing I have said, I repeat, should be taken as justifying discrimination; the real question is what part of the severe problems of the black population can be reached by programs of preferential hiring or is the result of present discrimination in employment. (Perhaps all of it can be attributed to *past* discrimination in employment, but that does not mean these problems can be *presently* reached by programs of preferential employment.)

V

If the conditions of the black population can be improved by these programs, then undoubtedly that would be the best reason for them. For me, no consideration of principle—such as that merit should be rewarded, or that governmental programs should not discriminate on grounds of race or ethnic group—would stand in the way of a program of preferential hiring if it made some substantial progress in reducing the severe problems of the low-income black population and of the inner cities. Because I have doubts as to what this contribution will be, I take more seriously the third objection I have raised to preferential hiring: the creation of fixed ethnic-racial categories, the danger of freezing them, and the danger of their spreading.

The Department of Labor, apparently, was the organization

which decided that the "affected" or "protected" classes should consist of Negroes, Spanish-surnamed Americans, Native Americans, and Orientals.[39] It is a strange mix. Why just these and no others? We understand why Negroes and American Indians —they have been the subjects of state discrimination, and the latter group has been, in a sense, a ward of the state. Puerto Ricans, perhaps, are included because we conquered them and are responsible for them. We did not conquer most of the Mexican Americans. They came as immigrants, and why they should be "protected" more than other minorities is an interesting question. Other Spanish-surnamed Americans raise even more difficult questions. Why Cubans? They have already received substantial assistance in immigration and have made as much progress as any immigrant might expect. Why immigrants from Latin America, aside from Puerto Ricans, who must also be included among the "protected," "assisted"—and, of course, therefore counted—classes? Why Oriental Americans? They have indeed been subject in the past to savage official discrimination, but that is in the past. Having done passably well under discrimination, and much better since discrimination was radically reduced, it is not clear why the government came rushing in to include them in "affirmative action"—unless it was under the vague notion that any race aside from the white *must* be the victim of discrimination in the United States.

We could go on. Why Spanish immigrants? In what sense have they been treated worse than immigrants from Italy or Greece? And why (if they are included among the Spanish-surnamed) Sephardic Jews? Why not the Portuguese in New England, and the French Canadians? And why not dark-skinned immigrants from India, who are now a very substantial part of the new immigration?

It is perhaps easier to understand who is included—even though there are anomalies—than to understand who is excluded. Another set of guidelines ominously states:

Members of various religious and ethnic groups, primarily but not exclusively of Eastern, Middle, and Southern European ancestry, such as Jews, Catholics, Italians, Greeks, and Slavic groups, continue to be excluded from executive, middle-management and other job levels because of discrimination based upon their religion and/or national origin. These guidelines are intended to remedy such unfair treatment.[40]

Clearly, these guidelines have been issued less because of a powerful need or demand of these groups for redress (though what is asserted is undoubtedly true) than because either these groups must have decided they need protection for themselves owing to the preferred position that other groups are attaining or because equity seemed to demand of the guideline-setters that they be included, too. Thus groups that were once content to press their advance through education, business, informal pressures, and specific complaints because of discrimination to state and Federal agencies may find, in self-defense or the desire not to lose an advantage, that they, too, must enter the arena of conflict that the government has defined by creating what we must now call, I assume, "more protected" or "more affected" classes—ethnic groups number 1, as against 2, in terms of their claim to governmental consideration.

Thus the nation is by government action increasingly divided formally into racial and ethnic categories with differential rights. The Orwellian nightmare ". . . all animals are equal, but some animals are more equal than others, . . ." comes closer. Individuals find subtle pressures to make use of their group affiliation not necessarily because of any desire to be associated with a group but because groups become the basis for rights, and those who want to claim certain rights must do so as a member of an affected or protected class. New lines of conflict are created, by government action. New resentments are created; new turfs are to be protected; new angers arise; and one sees them on both sides of the line that divides protected and

75

affected from nonprotected and nonaffected. We should not underestimate the effects of government benefits. If people begin by feeling they do not deserve them, they will soon change their minds to decide they do after all (otherwise, why would the government give it to them?) and angrily rise up to protect them.

Conceivably, there have been benefits as we have moved from nondiscrimination to soft affirmative action to harder goals and deadlines. But there have been losses, too.

CHAPTER

3

Affirmative Action in Education: The Issue of Busing

I

IT IS THE FATE of any social reform in the United States—perhaps anywhere—that, instituted by enthusiasts, men of vision, politicians, statesmen, it is soon put into the keeping of full-time professionals. This has two consequences. On the one hand, the job is done well. The enthusiasts move on to new causes while the professionals continue working in the area of reform left behind by public attention. But there is a second consequence. The professionals, concentrating exclusively on their area of reform, may become more and more remote from public opinion and, indeed, from common sense. They end up at a point that seems perfectly logical and necessary to them—but which seems perfectly outrageous to almost everyone else. I have argued in Chapter 2 that this is what has happened to the great national effort to overcome discrimination in employment. I will argue in this chapter that this is what has happened to the great national effort to overcome the shame of segregated schools for black and white in the United States.

For ten years after the 1954 Supreme Court decision in *Brown*, little was done to desegregate the schools of the South. But professionals were at work on the problem. The NAACP Legal Defense Fund continued to bring case after case into court to circumvent the endless forms of resistance to a full and complete desegregation of the dual school systems of the South. The Federal courts, having started on this journey in 1954, became educated in all the techniques of subterfuge and evasion, and, in their methodical way, struck them down one by one. The Federal executive establishment, first reluctant to enter the battle of school desegregation, became more and more involved.

The critical moment came with the passage of the Civil Rights Act in 1964, in the wake of the assassination of a President and the exposure on television of the violent lengths to which Southern government would go in denying Constitutional rights to Negroes. Under Title IV of the Civil Rights Act, the Department of Justice could bring suits against school districts maintaining segregation. Under Title VI, no Federal funds under any program were to go to districts that practiced segregation. With the passage of the Elementary and Secondary Education Act in 1965, which made substantial Federal funds available to schools, the club of Federal withdrawal of funds became effective. In the Departments of Justice and Health, Education and Welfare, bureaucracies rapidly grew up to enforce the law. Desegregation no longer progressed painfully from test case to test case, endlessly appealed. It moved rapidly as every school district in the South was required to comply with Federal requirements. HEW's guidelines for compliance steadily tightened, as the South roared and the North remained relatively indifferent. The Department of Justice, HEW, and the Federal courts moved in tandem. What the Federal guideline writers declared was adequate desegregation for purposes of receiving Federal funds was what the courts accepted as a legitimate specification of desegregation for their decisions. What the courts declared was what HEW and the Department of Justice Office for Civil Rights accepted in their turn as desegregation. Sometimes, the Executive branch was ahead of the Judicial; sometimes, the reverse.

The Federal government and its agencies were under continual attack by the civil rights organizations for an attitude of moderation in the enforcement of both court orders and legal requirements. Nevertheless, after 1964, there was an astonishing speeding-up in the process of desegregating the schools of the South.

Writing during the Presidential campaign of 1968, Gary Orfield, in his detailed study, *The Reconstruction of Southern Education,* stated:

To understand the magnitude of the social transformation in the South since 1964, that portrait of hate [of black students walking into Little Rock High School under the protection of paratroopers' bayonets] must be compared to a new image of tense but peaceful change. Even in the stagnant red clay counties in rural backwaters, where racial attitudes have not changed much for a century, dozens or even hundreds of black children have recently crossed rigid caste lines to enter white schools. Counties with well-attended Ku Klux Klan cross-burnings have seen the novel and amazing spectacle of Negro teachers instructing white classes. It has been a social transmutation more profound and rapid than any other in peacetime American history.

This is a revolution whose manifesto is a court decision and whose heroes are bureaucrats, judges, and civil-rights lawyers. . . .[1]

Mr. Orfield, however, thought that all this was coming to an end. With Nixon attacking the guidelines that had brought such progress, and with the civil rights coalition coming apart in the fires of the cities of the North, Mr. Orfield wrote, "A clear electoral verdict against racial reconciliation [that is, the election of Mr. Nixon] could mean that the episode of the school guidelines may recede into history as an interesting but futile experiment." [2] Mr. Orfield underestimated the bureaucrats, the courts, and the overall American commitment to the desegregation of Southern schools. The Nixon Administration did make a rather weak effort at the beginning to modify the march of desegregation as a concession to the South. Indeed, in early 1970 the Republican-appointed head of the Office for Civil Rights in HEW, Leon Panetta, was fired. Reading his account of his stewardship, one is amazed that he was not fired earlier: he refused to accept any consideration, political or otherwise, stemming from the White House or his appointed political superiors as to how he should carry on the work of his office, despite the fact that he was a political appointee. While he wrote his book to demonstrate that the Nixon Administration had given up the fight to desegregate the schools, one possible conclusion of his account is that the Nixon Administration had given up the effort to mollify important Southern senators because one determined official, the courts, and the entire Federal bureaucracy were just too much for the White House to take on. In his bureaucratic rigor, Panetta even denied nonconforming Southern school districts relief aid in the wake of Hurricane Camille, despite the intervention of Vice-President Agnew. There is much to be learned from this fascinating account about how one ideological and narrow-minded bureaucrat may stalemate a national administration. [3]

In any case, after 1969, when the Supreme Court, against the administration's effort at delay, ordered the immediate implementation of HEW desegregation plans in Mississippi, no further delay was allowed and desegregation in the South proceeded rapidly.

While Mr. Nixon's appointees continued to suffer abuse for insufficient zeal, the desegregation of the racially divided school systems of the South proceeded. Thus the new Director of the HEW Office for Civil Rights, J. Stanley Pottinger, could summarize some of the key statistics as of 1970 in the following terms:

When school opened in the fall of 1968, only 18 per cent of the 2.9 million Negro children in the Southern states attended schools which were predominantly white in their student enrollments. In the fall of 1970, that figure had more than doubled to 39 per cent . . . [and] the percentage of Negroes attending 100 per cent black schools dropped . . . from 68 per cent to 14 per cent. In 1968, almost no districts composed of majority Negro (and other minority) children were the subject of federal enforcement action. It was thought . . . that the limited resources of government ought to be focused primarily on the districts which had a majority of white pupils, where the greatest educational gains might be made, and where actual desegregation was not as likely to induce white pupils to flee the system. . . . Forty per cent of all the Negro children in the South live in [such] systems. . . . Obviously, the greater the amount of desegregation in majority black districts, the fewer will be the number of black children . . . who will be counted as "desegregated" under a standard which measures only those minority children · who attend majority white schools.

In order to account for this recent anomaly, HEW has begun to extract from its figures the number of minority children who live in mostly white districts and who attend mostly white schools. Last year, approximately 54 per cent of the Negro children in the South who live in such districts attended majority white schools. Conversely, nearly 40 per cent of the 2.3 million white children who live in mostly black (or minority) districts, now attend mostly black (or minority) schools.[4]

There has been further progress since, and, if one uses as the measure the number of blacks going to schools with a majority of white children, the South is now considerably more integrated than the North.

Despite this remarkable progress, the desegregation of schools remains the most divisive of American domestic issues. Two large points of view can be discerned as to how this has happened. To the reformers and professionals who have fought this hard fight—the civil rights lawyers, the civil rights organizations, the government officials, the judges—the fight is far from over, and even to review the statistics of change may seem an act of treason in the war against evil. Indeed, if one is to take committed supporters of civil rights at their word, there is nothing to celebrate. In 1970, the Civil Rights Commission attacked the government in a massive report on the civil rights enforcement effort. "Measured by a realistic standard of results, progress in ending inequity has been disappointing. . . . In many areas in which civil-rights laws afford pervasive legal protection —education, employment, housing—discrimination persists, and the goal of equal opportunity is far from achievement." And the report sums up the gloomy picture of Southern school segregation, sixteen years after *Brown:* "Despite some progress in Southern school desegregation . . . a substantial majority of black school children in the South still attend segregated schools." [5] Apparently this measure of success was attained, at least for majority white districts (for majority black districts, it would be a statistical impossibility) in 1970, but in CRC's follow-up report, issued one year later, this was not even mentioned. The civil rights enforcement effort in elementary and secondary schools, given a low "marginal" score for November 1970 (out of four possibilities, "poor," "marginal," "adequate," and "good"), is shown as having regressed to an even *lower* "marginal" score by May 1971, after HEW's most successful year in advancing school integration!

It is true that as desegregation as such in the South receded as a focus of attention in the early 1970s, a second generation of problems came increasingly to the fore: dismissal or demotion of black school principals and teachers as integration progressed; expulsions of black students for disciplinary reasons; the use of provocative symbols (the Confederate flag, the singing of "Dixie"); segregation within individual schools based on tests and ability grouping; and the rise of private schools in which whites could escape desegregation. Further, alongside these new issues, there was the reality that the blacks of the North and West were also to some degree segregated, not to mention the Puerto Ricans, Mexican Americans, and others. The civil rights professionals thus could see an enormous agenda of desegregation before them and could not pause to consider a success which was already in their minds paltry and inconclusive. The struggle must still be fought, as bitterly as ever.

There is a second point of view as to why desegregation, despite its apparent success, is no success. This is the former Southern point of view, and now, increasingly, a Northern point of view. It argues that a legitimate, moral, and constitutional effort to eliminate the unconstitutional separation of the races (most Southerners now agree with this judgment of *Brown*) has been turned into something else—an intrusive, costly, painful, and futile effort to stabilize proportions of races in the schools through transportation. The Southern congressmen, who for so long tried to get others to listen to their complaints, now watch with grim satisfaction the agonies of Northern congressmen faced with the crisis of mandatory, court-imposed transportation for desegregation. On the night of November 4, 1971, as a desperate House passed amendment after amendment in a futile effort to stop busing, Congressman Edwards of Alabama said:

Mr. Chairman, this will come as a shock to some of my colleagues. I am opposing this amendment. I will tell you why. I look at it from a rather cold standpoint. We are busing all over the First District of Alabama, as far as you can imagine. Buses are everywhere. . . . People say to me, "How in the world are we ever going to stop this madness?" I say, "It will stop the day it starts taking place across the country, in the North, in the East, in the West, and yes, even in Michigan."

One of the amendments had been offered by Michigan congressmen, long-time supporters of desegregation, because what had been decreed for Charlotte, North Carolina, and Mobile, Alabama, had been decreed for Detroit and its suburbs.

As massive waves of antagonism to transportation for desegregation rise up in one threatened area after another, the liberal congressmen who have for so long fought for desegregation ask themselves whether there is any third point of view: whether they must join with the activists who say that the struggle is endless and they must not flag, even now; or whether they must take the position the Southerners fought for, and lost. To stand with the courts in their current line of decisions is, for liberal congressmen, political suicide. A Gallup survey in October 1971 revealed that 76 percent of respondents opposed busing, almost as many in the East (71 percent), Midwest (77 percent), and West (72 percent) as in the South (82 percent); a majority of Muskie supporters (65 percent) as well as a majority of Nixon supporters (85 percent). Even more blacks opposed busing than supported it (47 to 45 percent). That is more or less how opinion stands today. But while to support the extension to all the Northern cities and suburbs of transportation for desegregation is suicide, how can liberal congressmen join with what they view as bigotry in opposing busing? Is there a third position, something which responds to the wave of frustration at court orders, and which does not mean the abandonment of hope for an integrated society?

How have we come from a great national effort to repair a monstrous wrong to a situation in which the sense of right of great majorities is offended by policies which seem continuous with that once noble effort? In order to answer this question, it is necessary to be clear on how the Southern issue became a national issue.

I I

After the passage of the Civil Rights Act of 1964, the first attempt of the South to respond to the massive Federal effort to impose desegregation upon it was "freedom of choice." There still existed the black schools and the white schools of a dual school system. But now whites could go to black schools (none did) and the blacks could go to white schools (few dared). It was perfectly clear that, in most of the South, "freedom of choice" was a means of maintaining the dual school system. In 1966, HEW began the process of demanding statistical proof that substantially more blacks were going to school with whites each year. The screw was tightened regularly, by the courts and HEW, and finally, in 1968, the Supreme Court gave the *coup de grâce,* insisting that dual school systems be eliminated completely. There must henceforth be no identifiable black schools and white schools, only schools.

But one major issue remained as far as statistical desegregation was concerned: the large cities of the South. For the fact was that the degree of racial concentration in the big-city Southern schools was, by 1968, no longer simply attributable to the dual school systems that they, too, had once maintained; in some instances, indeed, these schools had even been "satisfactorily" (by some Federal or court standard) desegregated years before. What did it mean to say that their dual school systems must also be dismantled "forthwith"?

To make the contrast clear, consider the case of rural New Kent County in Virginia, where the Supreme Court declared, in 1968, that "freedom of choice" would not be accepted as a means of desegregating a dual school system. Blacks and whites lived throughout the county. There were two schools, the historic black school and the historic white school. Under "freedom of choice," some blacks attended the white school, and no whites attended the black school. There was a simple solution to desegregation, here and throughout the rural and small-town South, and the Supreme Court insisted in 1968, fourteen years after *Brown,* that the school systems adopt it: to draw a line which simply made two school districts, one for the former black school, and one for the former white school, and to require all children in one district, white and black, to attend the former black school, and all children in the other, white and black, to attend the former white school.

But what now of Charlotte, Mobile, Nashville, Norfolk? To draw geographical lines around the schools of these cities, which had been done, meant that many white schools remained all white, and many black schools remained all black. Some schools that had been "desegregated" in the past—that is, had experienced some mix of black and white—had already become "resegregated"—that is, become largely black or all black as a result of the population movements common to all large central cities in the nation.

If there were to be no black schools and no white schools in the city, one thing at least was necessary: massive transportation of the children to achieve a proper mix. There was no solution in the form of contiguous geographical zoning.

But if this was the case, in what way were the Southern cities different from Northern cities? In only one respect: The Southern cities had once had dual school systems, and the Northern cities had not. (Even this was not necessarily a decisive difference, for Indiana had a law permitting them until the late 1940s, and other cities had maintained dual systems somewhat earlier.) Almost everything else was the same. The dynamics of population change were the same. Blacks moved into the central city; whites moved out to the suburbs. Blacks were concentrated in certain areas, owing to a mixture of formal or informal residential discrimination, past or present, economic incapacity, and taste; and these areas of black population became larger and larger, making full desegregation by contiguous geographical zoning impossible. Even the political structures of Southern and Northern cities were becoming more alike. Southern blacks were voting, liberal candidates appealed to them, Southern blacks sat on city councils and school boards. If one required the full desegregation of Southern cities by busing, then why should one not require the full desegregation of Northern cities by busing?

Busing has often been denounced as a false issue. Until busing was decreed for the desegregation of Southern cities, it was. As has been pointed out again and again, buses in the South regularly carried black children past white schools to black schools, and white children past black schools to white schools. When "freedom of choice" failed to achieve desegregation and geographical zoning was imposed, busing sometimes actually declined. In any case, when the school systems were no longer allowed to have buses for blacks and buses for whites, certainly the busing system became more efficient. Until 1971, busing for desegregation simply replaced busing for segregation.

But this was not true when busing came to Charlotte, North Carolina, and many other cities of the South in 1971, after the key Supreme Court decision in *Swann* v. *Charlotte-Mecklenburg County Board of Education*. The city of Charlotte is sixty-four square miles, larger than Washington, D.C., but it is a part of Mecklenburg County, with which it forms a single school district of 550 square miles, which is almost twice the size of New York City. Many other Southern cities (Mobile, Nashville, Tampa) also form part of exceptionally large school districts. While 29 percent of the schoolchildren of Mecklenburg County were black, almost all lived in Charlotte. Owing to the size of the county, 24,000 of 84,500 children were bused for the purpose of getting children to schools beyond walking distance. School zones were formed geographically, and the issue was: Could all-black and all-white schools exist in Mecklenburg County, if a principle of neighborhood school districting meant they would be so constituted?

The Supreme Court ruled they could not, and transportation could be used to eliminate black and white schools. The Court did not argue that there was a segregative intent in the creation of geographical zones—or that there was not—and referred to only one piece of evidence suggesting an effort to maintain segregation: free transfer. There are situations in which free transfer is used by white children to get out of mostly black schools, but if this had been the problem, the Court could have required majority-to-minority transfer only (in which one can only transfer from a school in which one's race is a majority, and to a school in which one's race is a minority), as is often stipulated in desegregation plans. Instead the Court approved a plan which involved the busing of some 20,000 additional children, some for distances of up to fifteen miles—from the center of the city to the outer limits of the county, and vice versa.

When this happened in Charlotte and, following this decision, in many other Southern cities, then busing was no longer a false issue. Busing for integration did not replace busing for segregation: A new school pattern was created by judicial order. The cost of busing in the Charlotte-Mecklenburg school district quadrupled between 1968–69 and 1969–70; the bus fleet had to be doubled from 267 to 516 at a cost of $1,500,000; and the number of children bused rose to 47,000, from 27 percent to 61 percent of the children in the school district.[6] Nor did it help to say busing was not the issue because so many American schoolchildren go to school by bus anyway. They go to school by bus either because the nearest school is far away and that is the only way to get there or because their parents prefer another school for them. This is radically different from busing for desegregation, when children are assigned to distant schools even though schools their parents prefer are nearer, and under compulsion, rather than as a result of a free parental choice.

This key Charlotte decision had important implications for Northern and Western cities without a history of legally mandated dual school systems.

Two implications of the decision remained then and still remain uncertain. If Charlotte, because it is part of the geographically large school district of Mecklenburg County, can be racially balanced, with each school having a roughly 71–29 white-black proportion, should not city boundaries be disregarded in other places and huge school districts of the Mecklenburg County scale be created wherever such action would make a better balance possible? A Federal district judge subsequently ruled that Richmond should be combined with suburban school districts to form a larger school district for purposes of creating a district in which black children might be in a minority. A Federal district judge ruled similarly in Detroit, requiring the creation of an even larger district. This case reached the Supreme Court, which sent it back with an ambiguous ruling: Unconstitutional segregation had been proved in Detroit, but not in the suburban districts. But the rules for proof of unconstitutional segregation are now so loose, as we shall show below, that it is hardly conceivable under present judicial standards that one cannot prove unconstitutional segregation in the Detroit suburbs, too.

There is a second implication of the Charlotte decision which has direct bearing on every Northern and Western city: If Charlotte is—except for the background of a dual school system—socially similar to many Northern cities, and if radical measures can be prescribed to change the pattern that exists in Charlotte, should they not also be prescribed in the North? And to that question, Federal judges ruling in San Francisco, Denver, Boston, and elsewhere have returned an affirmative answer.

The final resolution of this issue also still remains uncertain because courts—and the Supreme Court—have not yet said that one may impose transportation for desegregation in order to overcome the common cause of segregation in both Northern and Southern cities, residential concentrations of the two races. Rather, the legal position is that the latter situation—racial concentrations of the two races leading to racial concentrations in the schools—is simply "de facto segregation" and not unconstitutional. It is only when we have "de jure segregation"—a concentration created by public action to segregate—that segregation is unconstitutional, and judicial remedies, such as transportation for deconcentration, may be imposed. However, the standards for proof of "de jure segregation," as we shall show, have become so bent that it is scarcely conceivable under the present judicial standards that any situation of "de facto segregation" cannot be shown to be "de jure." The reason one can say that the two issues—of requiring school district mergers for deconcentration and requiring transportation to overcome school concentrations that are the result of residential concentrations—are still unsettled is that the Supreme Court may still withdraw from the advanced positions it has taken. These positions make it easy to require both school district merger and transportation to produce racial balance because they make it easy to prove racial imbalance is caused by state action.

How a "de facto" situation can be labeled "de jure" may be demonstrated from the case of San Francisco. San Francisco probably had a larger measure of integration than most Northern cities, but, de facto segregation had long been an issue in San Francisco. The school board proposed to set up two new integrated complexes, using transportation to integrate, one north and one south of Golden Gate Park. They were to open in 1970. When, however, one was postponed because of money problems, suit was brought charging de jure segregation on the ground that the school board's *failure* to implement the two integrated school complexes amounted to an official act maintaining the schools in a segregated state.

Judge Stanley Weigel, before whom the matter was argued, determined, on the basis of this and other findings, that what had previously been considered de facto was de jure: "The law is settled," he declared, "that school authorities violate the constitutional rights of children by establishing school attendance boundary lines knowing that the result is to continue or increase substantial racial imbalance."

Now, one can well imagine that a school board, even in the absence of state laws in the past that required or permitted segregation, could nevertheless through covert acts—which are equally acts under state authority—foster segregation. It could, for example, change school zone lines, so as to confine black children in one school and permit white children to go to another. It could build or expand schools so as to contain an all-black or all-white population. It could permit a transfer policy whereby white children could escape from black schools while blacks could not. It could assign black teachers to black schools and white teachers to white schools.

Judge Weigel charged (or, as the legal language goes, "found") all these things. The record—a record made by a liberal school board, appointed by a liberal mayor, in a liberal city, with a black president—does not, in this layman's opinion, bear him out, unless one is to argue that any action of a school board in construction policy, or zone-setting, or teacher assignment that precedes in time a situation in which there are some almost all-black schools (there were no all-black schools in San Francisco) and some almost all-white schools (there were no all-white schools in San Francisco) can be considered de jure segregation.

Under Judge Weigel's interpretation, there was no such thing as de facto segregation. All racial imbalance was the result of state actions, either taken or not taken. If not taken, they should have been taken. De facto disappeared as a category requiring any less action than de jure.

It turns out that Judge Weigel did go too far: Under the next ruling of the Supreme Court after *Swann* on Northern school desegregation, *Keyes* v. *School District No. 1, Denver,* in 1973, the distinction between de facto and de jure was still saved— barely; and in addition, the Court declared that some showing of intent must be made. On this basis, the Ninth Circuit, without undoing the massive desegregation that had occurred on the basis of Judge Weigel's too cavalier dismissal of all distinctions between de jure and de facto, turned back the orders of the court for further consideration.

But under the standards of *Keyes,* it will take little legal ingenuity to transform whatever might be considered de facto in San Francisco into de jure. Denver had been charged with unconstitutional segregation because, as in the case of San Francisco, its school board had undertaken to reduce black concentrations in certain schools. An election produced a new majority which rescinded the actions of the previous board and substituted a voluntary student transfer program. Legal action was then started. The District Court found that in one area, through the building of a small school rather than a larger one, by gerrymandering school zones, by using "optional zones," by using mobile classrooms, the school board had engaged in unconstitutional (de jure) segregation. However, no such finding was made by the District Court in connection with the black schools of another part of Denver, the central city core. This the District Court considered de facto segregation. The Supreme Court ruled, however, that once one finds segregation and segregative intent in one part of a school system, it can be assumed that segregation in other parts of the system are *also* the result of segregative intent, and thus the entire system may be considered de jure segregated. Thus a small amount of segregation with intent will be considered enough to justify massive desegregation through transportation for an entire system.

How elastic the *Keyes* standard is may be seen from one of the more outlandish of the findings of Judge Garrity in the Boston school case in 1974 following *Keyes*. Boston has high schools to which entrance is by examination, as do New York and other cities. One of them, Boston Latin, is old and famous, and has devoted teachers and alumni. As in the case of other examination schools in other cities, relatively fewer black students than white and Chinese do well enough on the examinations to gain entrance. There was no evidence these entrance examinations in any way discriminated against blacks. Nevertheless, Judge Garrity ruled that since he had found intentional segregation in the Boston schools in other respects, under the *Keyes* standard the existence of "racial segregation" (that is, underrepresentation) in the examination schools establishes, in the language of the *Keyes* decision, ". . . a prima facie case of unlawful segregative design on the part of school authorities." Therefore, "The burden of disproving unlawful intent thereupon falls upon the defendants, who in this case failed to carry the burden." One does not know whether to be more amazed at Judge Garrity's reasoning or the lawyers for the School Committee, who were apparently incapable of clearing even the innocent and long-established examination schools from a finding of unconstitutional racial segregation.

The hardy band of civil rights lawyers now glimpses the possibility of a complete victory, based on the idea that there is no difference between de facto and de jure segregation, an idea which is itself based on the larger idea that there is no difference between North and South. What was imposed on the South must be imposed on the North.[7] As Ramsey Clark, a former Attorney General of the United States, has put it, echoing a widely shared view among civil rights lawyers:

In fact, there is no de facto segregation. All segregation reflects some past actions of our governments. The FHA itself required racially restrictive covenants until 1948. But, that aside, the consequences of segregated schooling are the same whatever the cause. Segregated schools are inherently unequal however they come to be and the law must prohibit them whatever the reason for their existence.

In other words, whatever exists is the result of state action. If what exists is wrong, state action must undo it. If segregated schools were not made so by official decisions directly affecting the schools, then they were made so by other official decisions—Clark, for example, as does Judge Garrity, points to an FHA policy in effect until 1948—that encouraged residential segregation.[8] Behind this argument rests the assumption, now part of the liberal creed, that racism in the North is different, if at all, from racism in the South only in being more hypocritical. All segregation arises from the same evil causes, and all segregation must be struck down. This is the position that many Federal judges are now taking in the North.

I I

I believe that three questions are critical in developing a position on school busing. First, do basic human rights, as guaranteed by the Constitution, require that the student population of every school be racially balanced according to some specified proportion, and that no school be permitted a black majority? Second, whether or not this is required by the Constitution, is it the only way to improve the education of black children? Third, whether or not this is required by the Constitution, and whether or not it improves the education of black children, is it the only way to improve relations between the races?

These questions are, in practice, closely linked. What the Court decides is constitutional is very much affected by what it thinks is good for the nation. If it thinks that the education of black children can only be improved in schools with black minorities, it will be very much inclined to see situations in which there are schools with black majorities as unconstitutional. If it thinks race relations can only be improved if all children attend schools which are racially balanced, it will be inclined to find constitutional a requirement to have racial balance.

This is not to say that the courts do not need any authority in the Constitution for what they decide. But this authority is broad indeed and it depends on a doctrine of judicial restraint —not characteristic of the Supreme Court and subordinate Federal courts in recent years—to limit judges in demanding what they think is right as well as what they believe to be within the Constitution. Indeed, it was, in part, because the Supreme Court believed that Negro children *were* being deprived educationally that it ruled as it did in *Brown*. They were being deprived because the schools were very far from "separate and equal." But even if they were "equal," their being "separate" would have been sufficient to make them unconstitutional: "To separate them from others of similar age and qualifications simply because of their race generates a feeling of inferiority as to their race and status in the community that may affect their hearts and minds in a way unlikely ever to be undone."

While much has been made of the point that the Supreme Court ruled in *Brown* as it did because of the evidence and views of social scientists as to the effects of segregation on the capacity of black children to learn, the major ground of the decision, it seems clear, was that distinctions by race should have no place in American law and public practice, neither in the schools, nor, as subsequent rulings asserted, in any other area, whether in bus stations or golf courses. This was clearly a matter of the "equal protection of the laws."

The constitutional question became more complicated when the test for findings of segregation and remedies for desegregation became a *statistical* distribution of black and white children in the schools, because then the courts had to decide what statistical distribution would qualify as "segregated" and "desegregated." In the South, this became crucial when state laws no longer existed assigning black pupils to black schools and white pupils to white schools, when various evasions had been struck down, and when what replaced it was a racially neutral system of pupil assignment by residence rather than race. In the North, where discriminatory state laws and a dual school system did not exist in 1954, the issue of what statistics might be taken as a showing of segregation or of desegregation was crucial from the beginning. When we moved from laws mandating segregation to practices that might be or might not be racially tainted, and whose test was statistical distributions, we left behind any general public agreement with the courts when they found unconstitutional segregation. We entered a murky area in which one side could appeal to the Constitution while the other could attack judicial dictatorship. If court decisions based on statistical pattern are to have the same moral authority as court decisions based on uncontested facts of state law and practice, then school concentration stemming from residential concentration—"de facto segregation"—must be seen as no different from segregation imposed by law. Are there real distinctions between these two conditions?

The terms themselves bias discussion in a certain direction: "segregation," after all, is "segregation." If we had used from the beginning—as we might well have done—"concentration" for de facto segregation, some of the points at issue would have emerged more sharply. "Segregation" implies a conscious act to segregate, as well as a condition in which one group is strictly demarcated and separated from another. It was appropriately used for the situation in the South. It was inappropriately used for the situation in the North, where a "segregated" school is considered any that has a higher proportion of blacks than the school district as a whole, even when that proportion is less than a majority (47 percent black was the magic number defining "segregation" in San Francisco), and even when that higher proportion consists of black teachers and principals in a system who would like to work with black children. "Concentration," a more neutral term, raises the question as to why there is concentration, and at least opens the possibility that many of the reasons might be legitimate and constitutionally innocent. But there is no point arguing with predominant usage: "de facto" segregation is what "concentration" is now called. To the government agencies and the courts, a higher concentration of blacks in a school than in a school district is called "de facto segregation."

But since we are also interested in the educational effects of these two conditions—which I have suggested affects what judges find and decree—we must ask: Is there really a meaningful difference educationally between a 100 percent black school under a law that prohibits blacks from going to school with whites and an *x* percent black school where there are no such requirements? Let us make the comparison even stricter. Suppose a school is 100 percent black in a system which has no state laws or regulations requiring separation between the races. I believe there is still a difference. In the de facto situation, not all schools are 100 percent segregated. A child's observation alone may demonstrate that there are many opportunities to attend integrated schools. The family may have an opportunity to move, the city may have open enrollment, it may have a voluntary city-to-suburb busing program. The child may conclude that, if one's parents wished, one could attend another school; or that one could if one lived in another neighborhood—not all are inaccessible economically or because of discrimination—or he could conclude that the presence of a few whites indicated that the school was not segregated.

Admittedly, social perception is a complicated thing. The child in a 100 percent black school as a result of residential concentration may see his situation as identical to that of a child in a 100 percent black school because of state law requiring separation of the races. But the fact is that a black child in a school 20 or 30 percent black may also see himself as unfairly deprived. Or any black child at all, in view of the nature of black history and the currently prevailing interpretation of that history—even if the child is the only black in a white school—may come to this conclusion. Perception is based not only on reality, but also on the picture of reality transmitted by the media, authorities, peers, and the like. Under these influences, the lovely campuses of the West Coast may be seen as "jails" which confine young people, and those incarcerated by courts for any crime can be seen as political prisoners. If we feel a perception is wrong, one of our duties is to try to correct it, rather than assume that the perception of being a victim must alone dictate the action to be taken. False perceptions are to be responded to sympathetically, but not as if they were true.

Socially and social psychologically, there was a radical distinction between Southern school segregation and Northern school concentrations in the 1950s and 1960s.[9] But whatever the distinction in the past, it has narrowed to insignificance in the large cities as we have been successful in overcoming the distinctive system of law and prejudice which bound the black in the South. There is thus increasingly less justification for one law for the South and one for the North. One response might have been to forge a solution suitable for the nation as a whole, a solution that took into account the steadily declining role of public action in creating concentrations of black students. Another solution was to extend the law that had been forged in the South on the basis of an undisputed showing of state action to the North, and to try to demonstrate that any concentration of blacks beyond an even distribution was owing to state action there too, that there was no difference between the pre-1954 South and the North.

How could this be done? Under the *Keyes* doctrine, as we pointed out, intent must be demonstrated in order to have a finding of unconstitutional segregation, but only a certain amount of intent is necessary; for if one can find intent somewhere in the system, the presumption is that racial concentrations elsewhere in the system are the result of official action with the intent to unconstitutionally segregate. This has strengthened lawyers in their search for actions taken by school boards that have affected the racial distribution of children in the schools, such as changing a school-zone boundary when blacks moved into an area to keep a school all or mostly black or another one all or mostly white. In districts with 100 or more schools and a long history, with perhaps scores of school-zone lines changed every year, it would be unlikely if one could not come up with some cases. One could find that as a black school became underutilized, whites were not zoned into it. Or as a white school became overcrowded, whites were not zoned out of it. One could argue that schools were constructed in certain places so as to increase the likelihood that they would have higher concentrations of blacks or whites. Or that mobile classrooms had been attached to one school or another in order to prevent, or with the effect of preventing, an integration that might have occurred as children were required to move from an overcrowded school to another.

The search by which de facto is turned into de jure involves elaborate historical investigation into the changing of school-zone boundaries, the positions taken by citizens at school board meetings, the views expressed by school board members, and the like, making the records of these cases marvels of research of enormous length and complexity. However, these masses of data never quite satisfactorily settle the issue of how to separate intent to segregate from all the other intentions which affect the attitudes of parents and the decisions of school boards. Indeed, sometimes intent to segregate is simply the other side of intent to integrate. Parents (white and black) protest that their school is becoming dominantly black owing to residential movements and that they want it to remain integrated. The school board will oblige, drawing new school zones to preserve its integrated character. By so doing, it increases the blacks in another school. Ergo, intent to segregate has been demonstrated.

Whether one builds small schools (as in Detroit) or large ones (as in San Francisco), it may be found—both judges did— that the size of the school was determined by a desire to contain the black population. The location of a school will be a "finding" that there was segregatory intent, even though the black school populations of Northern cities have been rising so rapidly, and the process of school-building is so slow, that it is not easy for segregatory intention to be easily demonstrated or realized. One of the schools cited in the San Francisco decision as designed to "segregate" *blacks* (64 percent black in 1964) had been cited as recently as 1967 in the Civil Rights Commission's report on *Racial Isolation in the Public Schools* as having been built in order to "segregate" *whites,* since it had opened in 1954 with a student body that was almost all white.

The motives of public authorities are never unmixed. They are responsive to various constituencies, and in the case of school boards one of their chief constituencies is parents. *Their* motives are not unmixed, either. It seems to be the state of the law that whatever the character of any other intent—convenience, safety, responding to a constituency's desires for a school —if a school is disproportionately white or black, segregatory intent will be accepted by many judges. Thus, Judge Weigel in San Francisco accepted as evidence of de jure segregation the building of a new school in Hunter's Point, a black area. The school authorities had resisted building there, but the local people insisted on a new school. Just about everyone who supports desegregation in San Francisco supported the local people, even though they knew that the school would be segregated. The local NAACP also supported the building of the new school. Despite this, the judge, in his decision on the suit brought by the NAACP, cited the building of this school as a sign of the "segregatory" policies of the San Francisco school authorities. To the judge, the black people of Hunter's Point were being "contained," when they should have been sent off elsewhere, leaving their own area devoid of schools (or perhaps any other facilities). But in the view of the people of the area who demanded the school, they were being served. It would appear that the dominant intent of the school board was to respond to that demand. That did not save it.

In Boston, the school board opened a new school in a black section. It tried to save the state aid that would be lost if it did not take some action to desegregate, and it zoned white children living some distance away into the new school. Black parents living nearby insisted on enrolling their children in the brand new school. The white parents living further away also protested, and eventually the board succumbed to their pressure and allowed them to send their children to their old nearby schools. To the minds of most enforcers of school desegregation, state and national, and to Judge Garrity, the board condemned itself for a segregatory act. One of the things the boycotting parents said was that they were afraid their children would get beaten up going through the area they had to traverse in order to get to school. Who is to say that this was pure fantasy, given the conditions of the modern city, and that what the white parents really meant was that they did not want their children to go to a mostly black school? [10] But if a school board responds to such a concern, de facto is turned into de jure segregation.

One special twist in turning de facto school segregation into de jure involves the obligations of school boards. If the segregation is de facto, presumably there is no obligation in the school board to modify the racial distributions in the school. But suppose there is a state law (as in Massachusetts) requiring school boards to take action to reduce segregation, or there are state education regulations (as in New York, Pennsylvania, California, and other states) to the same effect, and that a school board does not take such action or resists proposals or orders by state authorities to take action: Then de facto segregation becomes de jure because the school board did not take action to reduce it. Or suppose the school board plans to take some action to reduce segregation but then, owing to whatever reason (the San Francisco board claimed lack of funds, but it might well be parental resistance), the action is rescinded. Then *that* rescinding of an action that was initially not constitutionally required will serve to transform de facto into de jure segregation.

Pro-busing forces may try, by a process that can only be called entrapment, to create such a situation. A liberal board on the way out will vote a plan that it knows its successor board will not accept (the liberal board might not have voted it either if it had had to implement it): When the conservative successor board rescinds the action we have de jure unconstitutional segregation.

Two examples will suffice:

Cincinnati's lame-duck school board voted a sweeping plan last night for the desegregation of the city's public schools, three weeks before a newly elected board pledged to resist integration takes office.

Taking the board's three conservatives by surprise, the four liberals on the board introduced a 14-page desegregation resolution without giving formal advance notice.

One of the conservatives left the room in protest and the resolution passed on a 4-to-2 vote.

It is expected that the first business of the new school board, which will have a 5-to-2 conservative majority, will be to rescind the lame duck board's action.

But the liberals, citing previous court decisions, say it is unlikely that the courts would permit them to do that.

The desegregation plan that was adopted represents the fruition of five years of struggle by Cincinnati liberals and Negro organizations.

It will completely redefine existing school attendance zones and require the busing of students to insure that an equal proportion of blacks and poor whites attend each of the city's schools.[11]

The same kind of tactic is reported in a study of school desegregation in the Richmond school district of California. There, after a liberal school board was overwhelmingly voted out of office, they strengthened a school desegregation measure so as to make possible a lawsuit against their conservative successors when they came into office and reversed the decision. The board president asserted, "We have changed the school boundaries to reflect an integrated district, and any attempt to change it back will wind up in a lawsuit." And the leader of the citizen group supporting the liberal members is quoted in this study as saying, ". . . the board redrew the boundaries at our insistence to lay the basis for a lawsuit." [12]

III

In these and other strange ways a legal finding of segregation can be made, and de facto can be changed into de jure; and when one considers that in a school board many motives may be present and may be found, it is scarcely possible that any situation in which there are varying concentrations of black children in the schools may not be found to be unconstitutional. On the basis of such a finding, massive remedies may then be imposed, and the only remedy in effect that will ensure an even distribution of the races in the schools in large cities is school assignment on the basis of race and compulsory transportation of children to the schools to which they are assigned, which will in many cases be distant from where they live. These remedies raise as many constitutional questions as the original findings.

Perhaps the first issue that is raised is the disproportionality between the segregation found and the scale of the remedy imposed. As we have seen from the *Keyes* decision, if there is a finding of some segregation in one part of a district, then a court may conclude that any racial concentration is the result of intentional segregation. It seems to be the state of the law as interpreted by the Federal judiciary that if a segregatory intent plays any part in school decisions, then any measure of relief, no matter how extensive, is justified. The relief favored by the courts and successful litigants is the dispersion of the black students, and of the students of other definable racial and ethnic groups (Spanish-surnamed and, Asian American in cases already decided, and perhaps others to come in cases not yet decided). Up to now the scale of remedy in cases has not been successfully challenged.

Perhaps the most serious issue raised by this kind of remedy is that it makes all but impossible one kind of organization that a democratic society may wish to choose for its schools: one in which the schools are the expression of a geographically defined community of small scale and regulated in accordance with the democratically expressed views of that community. We have had a good deal of discussion in recent years of "decentralization," "community control," and "parental control" of schools. There were reasons for "community control" long before the issue exploded in New York in the late 1960s, and there were reasons for "parental control" long before the educational voucher scheme was proposed. The current state of the law makes the school ever more distant from the community in which it is located and from the parents who send their children to it.

In Charlotte, as a result of the school busing, PTA activities declined and almost disappeared in some schools.[13] In San Francisco's Mission district, owing to the effective work of the Mission Coalition (an Alinsky-style community organization), the local community had considerable influence on public programs. With a wide base of membership, this organization could help determine what was most effective in the local schools. But if it wanted to create an atmosphere in the school best suited to the education of Spanish-speaking children, what sense does this make when the local schools are filled with non-Spanish-speaking children from distant areas? And how could it influence the education the Mission children received in the distant schools to which many of them were now sent? In Boston, as in San Francisco, bilingual education has become more difficult owing to dispersion.

By breaking the link between a community and its schools, the new decisions in effect shift power into the hands of the school bureaucracy, now constrained only by the Federal court and its experts, and released from any direct influence from parents too confused to know where to bring their protests. This is an ideal recipe for the ills of a mass society, in which it becomes obscure as to where responsibility is lodged, and discontent can find expression only by attacking large, distant, and presumably powerful objects, pointed out to it by some ideology.

In Boston, the elected school committee has been thrust aside, and the school bureaucracy negotiates, under Judge Garrity's overall direction, with his appointed masters and experts, who generally do not even live in Boston. All this serves to reduce the influence of people over their own lives and their own fates. The sense of powerlessness is accentuated, owing to the differential impact of school busing on rich and poor. The prosperous can avail themselves of private schooling, or they can "flee" to the suburbs. Even if metropolitan-wide districts are created under court order, making it impossible to "escape" by going to the suburbs, the class character of the decisions will remain or become even more pronounced. For while many working-class and lower-middle-class people can afford to live in suburbs, very few can afford the costs of private education.

Some observers have pointed out that leading advocates of transportation for integration—journalists, political figures, and judges—send their children to private schools which escape the consequences of these legal decisions. This does raise a moral question. The judges who impose such decisions, the lawyers who argue for them (including brilliant young lawyers from the best law schools employed by Federal poverty funds to do the arguing), would not themselves send their children to the schools to which, by their actions, others poorer and less mobile than they are must send their children. Those not subject to a certain condition are insisting that others submit themselves to it, which offends the basic rule of morality in both the Jewish and Christian traditions. One assumes there must be a place for this rule in the Constitution.

Thus one effect of the current crisis is that people already reduced to frustration by their inability to affect a complex society and a government moving in ways many of them find incomprehensible and undesirable must now see one of the last areas of local influence taken from them in order to achieve a single goal, that of racial balance. All people, black and white, should have the right to control as much of their lives as is possible in this complex society, and the schools are very likely the only major function of government which would not suffer—and might even benefit—from a greater measure of local control. In education, there are few "economies of scale." [14]

The Federal judges are quite sure that any kind of concession by school authorities to the needs or desires of given ethnic communities which would interfere with the statistical distribution the judges find desirable is unconstitutional. Judge Roth in Detroit is critical of the blacks of that city for contributing to what he considers "segregation" by demanding black principals and teachers:

In the most realistic sense, if fault or blame is to be found it is that of the community as a whole, including of course the black components. We need not minimize the effect of the actions of federal, state, and local governmental officers and agencies . . . to observe that blacks, like ethnic groups in the past, have tended to separate from the larger group and associate together. The ghetto is a place of confinement and a place of refuge. There is enough blame for everyone to share.

We would all agree with Judge Roth that the ghetto must not be a place of confinement and that everything possible must be done to make it as easy for blacks to live where they wish as it is for anyone else. But why should it be the duty or the right of the Federal government and the Federal judiciary to destroy the ghetto as a place of refuge if that is what some blacks want? Judge Roth is trying to read into the Constitution the crude Americanizing and homogenizing which is certainly one part of the American experience, but which is just as certainly not the main way we in this country have responded to the facts of a multiethnic society. The doctrines to which Judge Roth lends his authority would deny not only to blacks but to any other group a right of refuge which is quite properly theirs in a multiethnic society built on democratic and pluralist principles.

For some time now blacks, Mexican Americans, and Puerto Ricans, as well as Chinese, Italians, Irish, Jews, and so on, have found the idea of a community attractive, whether as a place of refuge, or to maintain values which need some degree of concentration to survive, or because the concentration of a group in a given area increases its political power and its access to jobs and influence. In Atlanta, there has been a strong black political presence which is now evident in the election of a black mayor and a black congressman. Local black leaders preferred not to press ahead for a full desegregation, with its attendant problems in the way of greater white flight, racial disturbances and tension, the loss of the black identity of certain schools of which the black community was proud, and the like. The local black leaders were ready to accept a settlement in which black schools remained and in which the number of black administrators in the school system was increased. This led to disagreement between the local chapter of the NAACP and the national NAACP, which wished to press ahead for full integration by whatever measures necessary, which, of course, would call for busing.[15] Local black leaders in Boston were also torn between the possibilities of greater community control and the busing plan, although almost universally they supported the latter.

Professor Derrick Bell, of the Harvard University Law School, who is black, recognizes the anomaly when he sets as a question in his casebook, titled *Race, Racism, and American Law*, the following:

Why are Northern Courts willing to "follow the law" in ordering school desegregation where white opposition is great, black support is minimal, and the applicability of Brown unsettled?

The direction of his thinking is perhaps suggested by the subsequent questions:

Is it significant in political and economic terms that, if blacks fail in the effort to require metropolitan school desegregation plans, they will soon assume control of the school system in many of the country's largest cities? . . .

With new social studies raising questions as to the value of integration in the educational achievement of black children, is it possible to argue that the authority to determine appointments of school principals is more important than the racial balancing of students in the schools? [16]

That the American constitutional order does not call for a total homogenization by force of all groups but rather the granting of individual and group freedom seems difficult for Federal judges to grasp. Judge Garrity again offers perhaps the most egregious example. In the Boston schools, most of the black teachers and principals are in the black schools. One reason is that the union contract gives teachers with seniority rights to transfer. Since black teachers have, on the whole, less seniority, more of them are in the majority black schools, where new teachers are most needed. But there is another reason. Judge Garrity writes:

Racial segregation [that is, concentration] of full-time faculty and staff has been attributable . . . to requests by principals and headmasters that black teachers be assigned to their districts. Such racial requests are received from ten to fifteen district principals a year and are honored by the personnel department wherever possible. . . .

Defendants have argued that assignment of black teachers to predominantly black schools is educationally justifiable because black teachers serve as adult role models and inspirational examples to black pupils. The record, however, is barren of evidence supporting this argument. One witness, associate superintendent Leftwich [he is black], testified that a need by black youngsters for adult role models would in his opinion be a logical explanation for granting requests of black teachers for assignment to black schools. But this was concededly a curbstone opinion unsupported by any study or empirical data.

Admittedly, the social sciences are not in such condition that decisive evidence supporting this common view could be presented, but one could quote 100 textbooks in education, psychology, and sociology that make more or less the same point.

It was to Judge Garrity inconceivable, apparently, that black principals and teachers might *want* to work in black majority schools; that they might have something distinctive to offer black children; that in a free society, living under our Constitution, this might be allowed. One thinks sadly of the work of a man like Leonard Covello, who worked for decades with Italian students in New York City because he wanted to improve the education of the children of his own community and who became a model of a teacher to be emulated, and realizes that in the world of the Judge Garritys and the Judge Roths this must not be allowed to happen: A community must not exist, and every distinguishable group must be distributed at random to ensure that it does not exist.

One understands that the people do not vote on what the Constitution means. The judges decide. But it is one thing for the Constitution to say that, despite how the majority feels, it must allow black children into the public schools of their choice; and it is quite another for the Constitution to say, in the words of its interpreters, that children, owing to their race or ethnic group alone, may not be allowed to attend the schools of their choice. When, starting with the first proposition, one ends up with the second, as one has in San Francisco and Boston, one wonders if the Constitution can possibly have been interpreted correctly.

Again and again, reading the briefs and the transcripts in these cases, one finds the words "escape" and "flee." The whites must not escape. They must not flee. Constitutional law often moves along strange and circuitous paths, but perhaps the strangest yet has been the one whereby, beginning with an effort to expand freedom—no black child shall be excluded from any public school because of his race—the law has ended up with as drastic a restriction of freedom as we have seen in this country in recent years: No child, of any race or group, may "escape" or "flee" the school to which that child has been assigned on the basis of his or her race, color, or national origin. No school district may facilitate such an escape. Nor may it fail to take action to close the loopholes permitting anyone to escape.

Even though the civil rights lawyers feel that in advocating the measures they now do they are in the direct line of *Brown,* something very peculiar has happened when the main import of an argument changes from an effort to expand freedom to an effort to restrict freedom. Admittedly the first effort concerned the freedom of blacks; the second, in large measure, concerns the freedom of whites (but not entirely, as we have seen from the resistance of the Chinese in San Francisco and many instances where blacks have resisted the elimination of black schools in the South and fought for community-controlled schools in the North). Nevertheless, the tone of civil rights cases has turned from one in which the main note is the expansion of freedom into one in which the main note is the imposition of restrictions. It is ironic to read in Judge Stanley Weigel's decision the following quotation from Judge Skelly Wright, written in those distant days when the fight was still for freedom:

The problem of changing a people's mores, particularly those with an emotional overlay, is not to be taken lightly. It is a problem which will require the utmost patience, understanding, generosity, and forbearance from all of us, of whatever race. But the principle is that we are, all of us, freeborn Americans, with a right to make our way, unfettered by sanctions imposed by man because of the work of God.

Apparently, Judge Weigel believed he was acting in the spirit of this quotation when he ordered the San Francisco School District to prepare a plan for the ". . . full integration of all public elementary schools so that the ratio of black children to white children will then be and thereafter continue to be substantially the same in each school, . . ." following which every child in the San Francisco elementary schools was placed in one of four ethnic or racial categories and made subject to transportation to provide an average mix of each in every school.

I V

One of the arguments justifying the assignment of children to schools by race and ethnic group is that this is a transitional measure to overcome racial discrimination and segregation. It is for this reason that certain provisions of the Civil Rights Act of 1964 have not restrained the courts.

Title IV of the Civil Rights Act ruled out school assignments for "racial balance":

. . . Nothing herein shall empower any official or court of the United States to issue any order seeking to achieve a racial balance in any school by requiring the transportation of pupils or students from one school to another or one school district to another in order to achieve such racial balance. (Title IV, 407 [a] [2])

The definitions in the title read, " 'Desegregation' means the assignment of students to public schools and within such schools without regard to their race, color, religion, or national origin, but 'desegregation' shall not mean the assignment of students to public schools in order to overcome racial imbalance" (401 [b]). This has become a dead letter for very much the same reasons as the similar prohibition on actions to correct imbalance in the employment title has become a dead letter: The courts and the Federal agencies (HEW and Justice) assert they are not requiring the assignment and transportation of students by race and so forth in order to achieve "racial balance," but in order to overcome "racial segregation," and that, of course, is the objective of the law, and the command of the Constitution. It is expected that at some point in the future, the direct command of the Constitution to public bodies to take no account of race, color, or national origin will come into effect.

The present requirement on all school districts to report their students, teachers, and administrators by race and ethnic group [17] is to be understood as based on the need to first determine whether there have been violations of the Constitution and the laws and then to fix a remedy. Racial and ethnic labeling is only a transitional feature. The question now becomes: How long does the transition last? The major process of the desegregation of Southern schools was complete five years ago, but I am not aware that any school district has ever been declared to be "unitary," so that it need no longer record the race and ethnic group of its children and is no longer subject to inquiry on its pupil-assignment procedures. Rather than solving the problem of the South, the new expansion of the meaning of desegregation has turned the entire country into actual and potential violators; nor is there any indication of the means by which this presumption of unconstitutionality and illegality may be lifted. Thus we see an exact parallel to the development of equal employment opportunity law. All, it is assumed, are guilty, and only the achievement of and maintaining of "representativeness" in employment or "racial balance" in school enrollments will get one off the hook. Nothing gets one off the hook permanently. For example, Charlotte-Mecklenburg has been busing for racial balance since 1971. It is still subject to the jurisdiction of the District Court. The chairman of the school board reports:

From 1969 to the present, Charlotte-Mecklenburg has used some version of every pupil assignment plan known to mankind, and even now we are trying to devise a new plan, so far unknown or undiscovered, which will give us that magic status known as "unitary." If that is achieved, our district judge avows—and he has said this several times publicly—that he will retire from the school board [sic]. He says that he doesn't know what a unitary school district is, but he thinks he will recognize one when he sees it.

The Supreme Court ruled on the Charlotte-Mecklenburg case in 1971 and said that it would not require a permanent process of school reassignment by race but only enough to overcome the vestiges of an unconstitutionally segregated dual school system. But consider the following account of what was happening in 1974:

Last summer we conducted a school assignment lottery because it was the only feasible way of complying with a court order which commanded us to find 600 white high school pupils living in the more affluent southeast section of our county to attend West Charlotte High School located in a large totally black neighborhood on the northwest perimeter of the city of Charlotte. The school board had proposed to send children from a white residential neighborhood much closer to West Charlotte, but the district judge refused to approve this plan. . . . The judge made it clear that he wanted socio-economic integration as well as racial integration. . . .

West Charlotte Senior High School . . . can house about 1,700 students. As of today 2,600 high school students live closer to this school than any other high school; 2,100 of them are black and 500 are white. To comply with the present court order, we must ship or bus out 1,500 black students and bring in some 600 white students selected on the basis of non-geographical criteria, and not only that, we must assure the court that the assignment plan is a stable plan—that is, one that cannot be circumvented, by white students in particular. We are presently operating under an order which forbids any change in assignment of a lottery-chosen white student assigned to West Charlotte High even if his family moves away from the area in which he was chosen. This has resulted in some bizarre transportation problems. In one case, we had to assign a driver and bus to pick up one pupil whose family built a new home and moved to the far northerly end of the county, more than 25 miles from West Charlotte High School.

One reason this is possible and why the case never comes to an end is that the plaintiffs can continue to raise new issues.

With the support of the government, the foundations, and the legal advocacy centers, this process can go on forever. For example:

> Proportionately, more black children have been bused—that is, taken beyond the nearest school and into a neighborhood of the opposite race—than have white children. The mathematics of racial balance, that is twice as many white children as black children, requires this in our district—yet, we are now told that this places an unfair burden on black children. Some white children have not been bused out but have remained at their home school to which black children have been brought. . . . The judge now calls these areas where white children attend their neighborhood school white sanctuaries and says they must be avoided in future pupil assignment plans.

Another reason the process can go on forever is that there are rapid changes in residence in this country (changes taking place for many reasons), and thus any racial balancing of the school on the basis of assigning pieces of territory to a school is unstable. No school ever opens with the racial balance that is predicted on the basis of a school census of the children in the territory assigned to it, since there are the alternatives of private schooling. So the chase goes on forever. Again from Charlotte-Mecklenburg:

> In several instances, after school assignment plans have been approved and the racial mix seemed to be in line, sudden and massive movements have taken place almost overnight. One development consisting of some 1,700 homes has changed from white to black in the last 2 years, and the change has all but wrecked the delicately balanced racial assignments in the affected schools.
>   . . . The court has demanded a stable school assignment plan —one which assures that all schools will remain predominantly white. Parents and children have also demanded a stable plan— one which lets them know where they can expect children to attend school from year to year. These two definitions of stability are obviously in conflict when viewed in the light of our recent history of rapidly changing residential areas and a highly mobile population.[18]

Presumably, there are circumstances—so the judge tells the school district—under which he would agree this search for perfect racial balance will come to an end and the school district will be free of the obligation to reassign children by race. But as soon as we try to envisage what those circumstances might be, we realize they can never come to pass. The requirement of the courts and the government is that a stable racial balance must prevail, for only that will give satisfactory evidence that segregation has been overcome. But that stable balance can be achieved only by a regular reassignment of pupils by race. If a school board should be considered to have finally reached desegregation and were to *stop* assigning children to schools by race, whether through a new neighborhood school plan, or freedom of choice, or education vouchers, or what you will, we may be sure that whatever racial distribution existed on the basis of assignment by race would change. Convenience, taste, and residential concentrations would lead to new racial concentrations in the schools. At that point, these new racial concentrations would be defined as coming into being through an act of "de jure segregation"—that is, the *stopping* of school assignment by race—and the process of "desegregation" would begin all over again. At some point, the courts and the government will have to recognize, if we are to be released from this *reductio ad absurdum,* that concentrations of students may come about for reasons aside from government action and need not necessarily be any sign of deprivation.

V

My discussion has concentrated on transportation for desegregation when it is required by the Federal courts. The desegregation guidelines of the Federal government would also entitle a Federal agency—the HEW Office for Civil Rights—to require such actions at the cost of loss of Federal funds, but since the great wave of Federal action to desegregate the schools in the South, the Federal government has moved cautiously, and it is generally the courts that have determined the difficult issues of when de facto becomes de jure segregation and what remedies may be required in such situations. Independently, the Department of Justice can require desegregation.

But quite independently of what the courts and the Federal government may require, local communities may undertake desegregation programs, including transportation, either because the local school boards feel they are good for education, or because they are under pressure from black and white liberal citizens who demand such measures, or because they are under pressure to do so from state authorities. Many such programs are voluntary and involve busing black children from central cities to suburbs. In Berkeley, California, and other cities children have been bused to achieve an even racial distribution in all the schools since 1968, without any court or Federal action.

I would make a sharp distinction between these actions by local communities and those mandated by courts. In the first case, we have a political process at work and the local school boards are responding to a demand from citizens, or to a requirement by state boards who wish to institute a certain educational policy. In the second, we have a requirement that may reflect no substantial local interest or demand, and in the face of which citizens, in their ordinary political roles, are simply helpless. Decisions to achieve racial balance taken by school boards not under judicial or Federal order but because the political forces in that district demand it will lead to less conflict than those undertaken under court order by resentful school administrations. In the first case, the methods of reducing racial imbalance have been worked out through the processes of political give-and-take, the community and teachers and administrators have been prepared for the change by the political process, the parents who oppose it have lost in what they themselves may consider a fair fight. The characteristics of judge-imposed decisions are quite different.

Admittedly, a strict adherence to a prohibition of the use of racial and ethnic categories in state action would ban these locally taken measures for more even distributions of schoolchildren by race as well as those required by Federal courts. Indeed, basing themselves on the Constitution and prohibitions of busing for racial balance, white parents have attacked various busing plans in the courts, charging that children were being assigned to schools for racial reasons alone. The courts have struck down all these challenges, making a distinction between legitimate state action on racial grounds to overcome discrimination and segregation, and illegitimate state action on racial grounds in order to discriminate and segregate. Since I prefer a greater measure of judicial restraint and a greater degree of freedom for democratically determined local action, I agree here with the courts: I would not like to see a community which wants to go further than others in creating even racial distributions hampered by a strict adherence to a ban on state action on the basis of race or ethnic group.

I would, however, like to see such distinctions made and used with great restraint. If one can substitute racially neutral territorial or income classifications for racial classifications, that would be all for the best, and we should move in that direction. Thus, a plan for busing might require not that *black* or *Spanish-surnamed* children be evenly distributed in the schools, but that children of *low achievement* be evenly distributed in the schools, or that children from *inner-city schools of low average achievement* be given the first opportunity to enter a program to bus children to suburban schools.

117

This would achieve the same aims, and would place the emphasis where it should now properly be, on issues of school achievement.

Nor would I exclude a role for findings that de facto segregation is, in reality, de jure segregation; that may be the case in many communities. What I would urge is that courts and Federal agencies become cognizant of all the other factors, aside from discrimination and segregation, that may lead to school concentrations of children and teachers: residential concentrations, a preference for neighborhood schools, the interests of teachers and administrators, the legitimate educational decisions of educators. It is the legal sophistry that attempts to define *all* racial and ethnic concentrations as de jure segregation that I criticize, not the legitimate effort which should be pursued by the Federal agencies and the courts to prevent school boards from segregating children and teachers by race and ethnic group where that is still going on.

Our test should be the expansion of freedom, within the limits of our overall national commitment to a society in which distinctions on grounds of race, color, and national origin are repugnant. If we accept that test, we will be saddened and depressed that one means of achieving a measure of desegregation through freely chosen acts by parents plays almost no role in court cases, and that is "freedom of choice" or "open enrollment." "Freedom of Choice" was given a bad name in the South when it was instituted as the least demanding response to the requirement to dismantle dual school systems, and when it was often combined with intimidation of black parents and children. But there have been many legitimate efforts to use freedom of choice to create a larger measure of integration. An examination of these efforts shows that, if they are administered effectively, substantial numbers of black parents will take advantage of them and that a greater measure of desegregation will be brought about than through racially neutral neighborhood school zoning.[19] Whatever else the conclusion of the constitutional debate on busing, we should all be able to agree that we should move as far as possible toward an even racial distribution through measures that permit free choice before we impose compulsion.

VI

I conclude there is considerable room for doubt as to whether the Constitution actually mandates a system whereby every school shall have a black minority and no school shall have a black majority. These policies may nevertheless be best for society. So we come to another crucial question raised by the new line of cases: Is an even racial balance necessary to improve the education of black children and the relations between the races?

Without rehearsing the terrible facts in detail, we know that blacks finish high school in the North three or more years behind whites in achievement. We also know with a fair degree of confidence that this huge gap is not caused by differential expenditures of money. Just about as much is spent on predominantly black schools outside the South as on predominantly white ones. Classes in black schools will often be smaller than classes in white schools—because the black schools tend to be located in old areas with many school buildings, while white schools tend to be in newer areas with fewer and more crowded school buildings. Blacks will often have more professional personnel assigned, owing to various Federal and other programs. There are, to be sure, lower teacher salaries in the predominantly black schools, because they usually have younger teachers with less seniority and fewer degrees. Anyone who believes this is a serious disadvantage for a teacher has a faith in experience and degrees which is justified by no known evidence. It is quite true that the big cities spend less on their schools, white and black, than the more prosperous surrounding suburbs, which are disproportionately white. Cities do spend more than small towns and rural areas. Regardless of the fact that spending more is unlikely to do much to improve education—it tends mostly to improve teachers' salaries and fringe benefits —it seems wrong that more public money should be spent on the education of those from prosperous backgrounds than on those from poorer families. But this is quite separate from the issue of whether, within present school districts, less is spent on the education of black children. It seems clear that in general this is not the case.

If money is not the decisive element in the gap between white and black, what is? In 1966, the Coleman report on *Equality of Educational Opportunity* reviewed the achievement of hundreds of thousands of American schoolchildren, black and white, and related it to social and economic background, to various factors within the schools, and to integration. In 1967, another study, *Racial Isolation in the Public Schools,* analyzed the effects of compensatory-education programs and reviewed the data on integration. Both studies—as well as subsequent experience and research—suggested that if anything could be counted on to affect the education of black children, it was integration. However, the operative element was not race but social class. The conclusion of the Coleman report still seems the best statement of the case:

. . . The apparent beneficial effect of a student body with a high proportion of white students comes not from racial composition per se, but from the better educational background and higher educational aspirations that are, on the average, found among white students.

On the other hand, if such integration did have an effect, it was not very great. The most intense re-analysis of Coleman's data concludes: [20]

Our findings on the school racial composition issue, then, are mixed . . . the initial *Equality of Educational Opportunity* survey overstressed the impact of school social class. . . . When the issue is probed at grade 6, a small independent effect of schools' racial composition appeared, but its significance for educational policy seems slight.

The study of these issues has reached a Talmudic complexity. The finding that integration of different socioeconomic groups favors the achievement of lower socioeconomic groups apparently stands up, but the effect is not large. One thing, however, does seem clear: Integrating the generally lower-income whites of the central city with lower-income blacks, particularly under conditions of resentment and conflict, as in San Francisco and Boston, is likely to achieve nothing, in educational terms. Furthermore, the effect of busing is to lose substantial parts of the white student population.

In San Francisco, the number of children enrolled in elementary schools dropped 6,519 against a projected drop of 1,508 (a 13 percent decline against a projected 3 percent decline) in response to Judge Weigel's decision. The junior-high-school enrollment, not yet subjected to full-scale busing, declined only 1 percent, and high-school enrollment remained the same. In Pasadena, California, there was a 22 percent drop in the number of white students in the school system between 1969—before court-imposed busing—and 1971. In Norfolk, Virginia, court-imposed busing brought a drop of 20 percent. In the massive confusion and conflict over the first year of busing in Boston, it was impossible to say how many children had left the system, but it was certainly large. Court-appointed desegregation planners and school planners differed wildly in their estimates of the number of children who were in, or potentially in, the school system in the spring of 1975, after a year of boycott, riot, and school closings. If, as seems reasonable, it is the somewhat better-off and more mobile who leave the public-school system when busing is imposed, as seems likely, the effect on the achievement of black children is further reduced.[21]

Nancy St. John has recently reviewed all the available studies —more than 100—on the effects of desegregation on children's educational achievement, on their self-image, and on racial attitudes. The findings are, in a word, complex. There is no simple relation between desegregation and any desirable outcome. It depends; as how could it not? [22] One would expect it to depend on the social class of the children, their family backgrounds, the way integration came, the conflicts surrounding it, the attitudes of their parents, the character of teachers and principals, the way administrators carried it out, and so on. To believe in any single-factor hypothesis—"desegregation will do thus and so"—is naïve.

It was in the light of such findings as Coleman's that judges in Detroit and Indianapolis and elsewhere called for combining the central city and the suburb into unified school districts. But if such an elaborate reorganization of the schools is to be undertaken so that the presumed achievement-raising effect of socioeconomic integration might occur, we are likely to be cruelly disappointed. There is little if any encouragement to be derived from studies, published and unpublished, of voluntary busing programs, even though such busing takes place under the most favorable circumstances (with motivated volunteers, from motivated families, and with schools acting freely and enthusiastically). Indeed, much integration through transportation has been so disappointing in terms of raising achievement that it may well lead to a reevaluation of the earlier research whose somewhat tenuous results raised what began to look like false hopes as to the educational effects of socioeconomic integration.[23]

Studies of the effect of integration have also considered

black-white attitudes and black self-image, and once again, the outcomes are complex.

But if one raises the larger question, isn't integration good for education? the answer must be, of course. It is desirable that children of various ethnic and racial groups should meet with each other and know each other. It will undoubtedly help their knowledge of other groups. Hopefully, stereotypes will be replaced by realities. Racist and liberal stereotypes alike are subject to this correction. If children from some group have attitudes or traits that hamper learning, they may learn better mixed in with children from another group (but on the other hand they may also be intimidated into apathy, or frustrated in their inadequacy, and resort to violence). The principle is not in question. But we often modify principles, good as they are, for other objectives. Thus, good as integration is, we permit Catholics and Jews to conduct religious schools for Catholics and Jews in which there is a loss from lack of contact with other children; but there is also a gain in the intensity with which a tradition can be taught and transmitted. The principle that public policy should strive toward an even distribution of children of different races and ethnic backgrounds in the schools is a good one, and it is my faith and belief—even if I cannot give chapter and verse from the educational research literature—that it is good, overall, for the education of all children. But even this principle may be modified by other principles. I would modify it to give some room for the principle of free parental choice.

I come to a similar conclusion when I consider the effect of a policy to achieve equal racial distributions in the schools by whatever measures necessary on relations between the races.

Of course, it is understood even by those who argue that busing will lead to better race relations that there will be a period of strain and disharmony when one introduces new mechanisms of school assignment and upsets old expectations, but it is expected that, if political figures and leaders accept the program that is necessary to provide majority white environments for black children, this difficult transition will be overcome and will be succeeded by a period of true integration in which the mixing of bodies by statistical formulas is replaced by better relations between people of different groups that are not affected by race or ethnicity.

There is much to this argument: Better relations between groups will not come about if they are rigidly separated. On the other hand, it may also be doubted that they will come about on the basis of mathematical formulas on how children from different groups should be mixed.

The experience of the city of Berkeley is instructive in this regard. It would be generally agreed that there are few communities which present as pleasing a picture of good race relations. The administration of Berkeley is in the hands of liberals, and thus it was possible for Berkeley to carry out, without court order or HEW requirement, or threat of either, a program in which each school would replicate the racial proportions of Berkeley as a whole. This program required busing and was not introduced without some conflict, but it has been the firm pattern of Berkeley school assignments since 1968. It is the largest city in the country voluntarily to have introduced a program of full racial balance through busing. The committee of Senator Mondale investigating equal educational opportunities in 1971 was naturally interested in Berkeley as a model of school integration. It discovered, not surprisingly, that the educational-achievement gap between black and white and Asian students had not particularly been affected. But certainly one could expect to find here a model of good race relations. It interviewed students from elementary schools, junior high schools, and the high school (there is only one in Berkeley and it has always been integrated, so it could not have been suffering from the consequences of transition to racial balance). The committee discovered, when it spoke to students—selected, one assumes, by the school authorities because they would give the best picture of integration—that students of different groups had little to do with one another. The black president of the senior class said:

. . . The only true existence of integration of Berkeley High is in the hallways when the bell rang everybody, you know, pass [sic] through the hallways, that is the only time I see true integration in Berkeley High.

Senator Brooke probed deeper. Since the young man was black and a majority of his classmates were white, had they not voted for him? "The whites didn't even participate in voting. . . . They felt the student government was a farce." (The opposing candidate was also black.) What about social activities? "Like we have dances, if there is a good turnout you see two or three whites at the dances. . . ." Intramural sports? "The basketball team is pretty integrated, the crew team is mainly white, soccer team mainly white, tennis team mainly white." Did this mean, Senator Brooke asked, ". . . that blacks don't go out for these teams that are white and whites don't go out for those teams that are all black?" The class president guessed that ". . . whites like to play tennis and blacks like to play basketball better." Still, he did think integration was a good idea, as did a Japanese girl who told the senators: "I think like the Asian kids at Berkeley High go around with Asian kids."

A Chicano student testified:

. . . I think now the Chicanos and blacks, they do hang around in groups. Usually some don't, I admit, like I myself hang around with all Chicanos but I am not prejudiced. I do it because I grew up with them, because they were my school buddies when there were segregated schools.

A white girl from the high school testified:

Integration, ideally, as far as I can see it isn't working. I mean like as far as everybody doing things together . . . I have one class where there are only two whites in it, I being one of them, you know, like I don't have any problems there, but outside . . . [with other blacks] we just do different things. I am not interested in games. I couldn't care less. I don't know anything about Berkeley as far as the athletics go. . . . I wear very short skirts and walking down the halls I get hassled enough by all the black Dukes, you know. . . .

Senator Brooke was surprised she wasn't hassled by the white boys, too, and suggested that they might use a different technique. The student witnesses were asked whether their parents had opposed integration: None had.[24]

This is about the most positive report one can make on school integration. Why should anyone be surprised? There is a good deal of hanging around in groups and there is some contact across racial lines, but the groups seem to have different interests and different social styles.

As anyone who has been in Berkeley knows, there are friendships across racial lines, as in other communities. And we might expect that as various groups lose distinctive traits and develop common attitudes that more friendships would naturally form across racial lines. But as any parent who has brought two children together because they "should be friends" knows, it is naïve to think that bringing groups physically together will introduce harmony into their relations. The processes of racial and group integration are rather more complex. Primarily they depend on common interests and tastes. It is understandable when distinctive subcultures, ethnic and occupational, have been created through a long historical process, that interests and tastes will not be distributed at random. Blacks will prefer one kind of music at a dance and whites another, one sport to another, the way different groups relate to people will vary, and the situations in which they are comfortable and at ease will vary. This is no argument against integration, an objective to which we are all committed. It is an argument against the hopes that the fixing of racial balance as the means by which every school system is to operate will help us move toward that goal.

VII

There is, unfortunately, a widespread feeling, strong among liberals who have fought so long against the evil of racial segregation, that to stop now—before busing and expanded school districts are imposed on every city in the country—would be to betray the struggle for an integrated society. They are quite wrong. They have been misled by the professionals and specialists—in this instance, the government officials, the civil rights lawyers, and the judges—as to what integration truly demands, and how it is coming about. Professionals and specialists inevitably overreach themselves, and there is no exception here.

It would be a terrible error to consider opposition to the recent judicial decisions on school integration as a betrayal of the promise of *Brown*. The promise of *Brown* was realized. Black children may not be denied admittance to any school on account of their race (except ironically in the cases in which courts and Federal officials insist that they are to be denied admittance to schools with a black majority simply because they are black). The school systems of the South are desegregated. But more than that, integration in general has made enormous advances since 1954. It has been advanced by the hundreds of thousands of blacks in Northern and Southern colleges. It has been advanced by the hundreds of thousands of blacks who have moved into professional and white-collar jobs in government, in the universities, in the school systems, in business. It has been advanced by the steady rise in black income which offers many blacks the opportunity to live in integrated areas. Most significantly, it has been advanced because millions of blacks now vote—in the South as well as the North—and because hundreds of blacks have been elected to school committees, city councils, state legislatures, the Congress. This is what is creating an integrated society in the United States.

We are far from this necessary and desirable goal. We can now foresee, within a reasonable time, the closing of many gaps between white and black. But I doubt that mandatory transportation of schoolchildren for even racial distribution will advance this process.

The increase in black political power means that blacks—like all other groups—can now negotiate, on the basis of and to the extent of their own power, about what kind of school systems should exist, and what measure of transportation and racial balance they involve. In the varied settings of American life, there will be many different answers to these questions. What Berkeley has done is not what New York City has done, and there is no reason why it should be. But everywhere black political power is present and contributing to the development of solutions.

There is a third path for liberals now agonized between the steady imposition of racial and ethnic group quotas on every school in the country—a path of pointlessly expensive and destructive homogenization—and surrender to segregation. It is a perfectly sound American path, one which assumes that groups are different and will have their own interests and orientations, but which insists that no one be penalized because of group membership and that a common base of experience be demanded of all Americans. It is the path that made possible the growth of the parochial schools, not as a challenge to a common American society, but as one variant within it. It is a path that, to my mind, legitimizes such developments as community control of schools and educational vouchers permitting the free choice of schools. There are as many problems in working out the details of this path as of the other two, but it has one thing to commend it as against the other two: It expands individual freedom, rather than restricts it.

One understands that the Constitution sets limits to the process of negotiation and bargaining even in a multiracial and multiethnic setting. But the judges have gone far beyond what the Constitution can reasonably be thought to allow or require in the operation of this complex process. The judges should now stand back and allow the forces of political democracy in a pluralist society to do their proper work.

CHAPTER

4

Affirmative Action in Housing: Overcoming Residential Segregation

I

EMPLOYMENT and education are closely linked; and both are linked to residential distribution. If parents in minority groups had better jobs with higher incomes, their children would undoubtedly do better in school. Alternatively, if the education received by minority-group children was better, they would go further in school and get better jobs. The connections here are not completely clear, and we are not certain what in higher-income families leads to their children doing better in schools, or what in doing better in school leads to better jobs. But we need not go into these presently much-disputed issues. The gross connections are clear enough, even if it turns out that the most effective factor in producing a better educational result for the children of the better-off is that the parents themselves have more education, or that the most effective factor in producing better jobs among those who get more schooling is that employers use the amount of schooling as a primary screening device for employment.

The connections among employment and education and residence in a metropolitan area are also close. Children in higher-income areas do better in school, and parents will go to great efforts to move into a community (or into a school zone) where a reputedly better school exists. In effect, a "better school" means a school with more children from middle-class, higher-income families, and a useful proxy parents use for this determination (though one that will be increasingly ineffective as black income rises) is the number of blacks in a school. The connection between residence and employment is looser and more disputable, but strong arguments have been made that one of the chief factors preventing a more rapid movement upward in black income and occupations is black concentration in the central cities of metropolitan areas. These central cities, the argument runs, lose jobs, while the suburban belts around the central cities gain jobs. Thus jobs move away from the black population, and transportation difficulties make it hard to search for the newer jobs in the growing parts of metropolitan areas, or to commute to them, lowering the occupations and incomes of blacks. There are other effects of this residential division between black and white. The major new building of houses and public facilities is in the suburbs; thus, the housing of blacks cannot improve as rapidly as it should and they use poorer public facilities. Furthermore, since there is a higher concentration of poor among the black, the concentration of the black is also a concentration of the poor, and this leads to something like a contagion of social problems. But perhaps the most important effect of inner-city concentration is on access to jobs, since industrial jobs are located where land is cheap and open, accessible to new thruways, and freer from congestion—the suburbs.

I

It is this argument and the policies designed to lead to a more even distribution of blacks (and other minority groups of low income) throughout metropolitan areas that I analyze in this chapter. These are some of the key facts that describe the situation.*

Twenty-one percent of central city families in 1969 had income under $5,000 a year as against 12 percent for the suburbs (but 27 percent for non-metropolitan areas). In 1970, 58 percent of blacks lived in central cities as against 27 percent of whites; conversely, 40 percent of whites lived in suburbs as against 16 percent of blacks. As a *percentage* of suburban population, blacks declined from 6.1 to 5.3 between 1950 and 1970; in central cities, they rose from 12.3 to 20.5 percent (blacks in the nation as a whole were 11 percent). Most of the new houses are built in the suburbs, in view of their more rapid growth. (But oddly enough, considerably more new houses are built in central cities as a ratio to population growth

---

* For the census, a "Standard Metropolitan Statistical Area" (SMSA) is a city of 50,000 with the county or counties around it, if they are related to it according to certain criteria. The census recognizes 243 such areas—most of which would not be considered "metropolitan" in ordinary parlance. The areas outside the central cities are generally called "suburbs"—though many would not be considered "suburban" in ordinary speech. Much of this area outside the central cities of SMSAs is farmland, or small towns, and alternatively, part of it may be smaller cities as "urbanized" as the central city. In 1970, 69 percent of the population lived in SMSAs. Central cities of SMSAs made up 31 percent of the entire population, suburbs 38 percent, the rest of the country 31 percent. (U.S. Bureau of the Census, *Census of Population and Housing: 1970, General Demographic Trends for Metropolitan Areas, 1960 to 1970, Final Report PHC* [2]–1 *United States,* Washington, D.C.: U.S. Government Printing Office, 1971.)

than in suburbs: "This means the degree of improvement in housing conditions was greater in [central cities] than in suburbs, in relation to population growth." [2] Presumably, it also contributed to the abandonment of housing.) And, to conclude this statistical presentation: Between 1960 and 1970 the central cities of the fifteen largest metropolitan areas *lost* 836,000 civilian jobs, while their suburbs *gained* 3,086,000 jobs.

As in the case of employment and education, we deal with a policy sequence in which blacks, and to a lesser degree other minorities, were discriminated against with impunity by landlords, homeowners, realtors, lenders, communities, and all others involved in the production, sale, or renting of housing. In the 1950s, states in the North began to pass Fair Housing laws and pressures developed for such a law at the national level. Simultaneously, discriminatory practices were struck down by the Federal courts. Thus, in *Shelley* v. *Kramer,* 1948, the Supreme Court ruled that a restrictive covenant—one preventing owners of property from selling it to blacks or members of other groups—could not be enforced in the courts. In 1962, by executive order, it was decreed that discrimination in renting or selling housing built under government subsidy or with the assistance of government-guaranteed mortgages would not be allowed. The Fair Housing Act of 1968 extended the ban on discrimination because of race, color, religion, or national origin to all housing except for a sale by the owners of individual homes or rental of a room or apartment in a house in which the renter lived. The ban extended to the actions of the real estate industry, to lenders of funds, to advertising. In housing and residential location, as in employment and education, the issue has now become how far beyond the ban on discrimination in the direction of "affirmative action" policy should extend.

We can discern the same general developments, but the differences among these large arenas of policy are also significant. The great majority of children in school attend public institutions. Public policy determines which school they can go to, and the potential sweep of affirmative action is massive. Employment is a matter of individual adult decision, but individual employers are often of enormous scale, and large institutions may be directed by public policy to take actions that will have a substantial effect on the pattern of employment. In both education and employment, there are significant levers of possible change and they are relatively few. In the case of housing, the matter is quite different. Only a tiny part of the existing housing stock was built with Federal subsidies. In the late 1960s and early 1970s, new housing policies rapidly increased the proportion of new housing built under Federal subsidies to mortgages to reach in some years as much as 20 percent of all new housing, but the legislation under which this housing was built has lapsed, and what part of new housing will be built with government subsidy is uncertain. A larger part of housing has been built with mortgages whose repayment has been guaranteed by the Federal government, but even this is only a small part of the housing stock. There are big builders of housing, but most housing is built by small builders and owned by individuals. We deal with a market of millions of individual purchasers and hundreds of thousands of individual suppliers. Discriminatory behavior may be policed, though that itself is difficult owing to the many subterfuges which are possible. Indeed, discriminatory practices seem substantially more widespread in the sale and rental of housing than in employment or education, though it is impossible to quantify this proportion in any of these areas.

Is it possible to end discrimination in the sale and rental of housing by moving beyond discrimination to affirmative action? And if one did, could one achieve a more even distribution of black and white between central city and suburb? In the field of housing and residential location, affirmative action raises more difficult problems in practice than in education and employment. School boards can be ordered to distribute children by race. Large employers can be ordered to hire persons of certain ethnic backgrounds in certain proportions. But how does one find an equivalent in housing? The purchase or rental of housing involves an individual choice limited by an economic capacity. One can require affirmative action to the extent of requiring large builders or renters to advertise the availability of their properties widely—and this is done now under law—but it would be difficult in a free society based on the market to say *so many* families of such and such a minority group must be recruited as sellers or renters.

Further, the fact that the average incomes of blacks are lower than those of whites means that a larger part of the housing market is out of their reach; for that reason alone, the ending of all discrimination and the achievement of some degree of affirmative action by large builders and renters would not necessarily change substantially the distribution of blacks between central cities and suburbs. And thus policy has moved to concentrate on a more significant obstacle to even distribution in residential locations: the power of local communities through zoning and other regulations on building to exclude low-income groups and, by implication, blacks. An extensive search for legal and political means to open up the suburbs to poor and black has become perhaps the most important thrust in the urban policy area. For example, in state after state, new housing corporations are being established; one of the most important issues affecting their powers is the extent to which they will be able to build or encourage building in the suburbs of subsidized housing, and the extent to which local communities will be able to retain power over zoning and other regulations that limit the powers of builders, public or private, to put in new housing. A complex legal strategy involving many participants is now being pursued and those engaged in it hope that some major court decision may emerge that sharply limits community powers and leads to a greater flow of the poor and the black into the suburbs. No other general strategy now competes

with the opening of the suburbs in its promise for an ameliora-
tion of so many urban problems. It is seen as the Gordian knot
of urban solutions.

The argument for opening up the suburbs raises issues in
three areas: of fact, of effect, and of policy. Are the *facts* as to
segregation such as to require major new strategies and policies
aside from a strict policing of the laws against discrimination?
Are the *effects* of a different distribution of black and poor in
metropolitan areas such as to require major new strategies and
policies? Are the *policies* proposed such that we may support
them with confidence in the hope that a better society will
follow?

II

Facts: Is the segregation of blacks, under the present state of
the law, and under the present and expected circumstances of
the distribution of income and jobs as between blacks and
whites, such that blacks are doomed to central city concentrations, regardless of their will? The crude answer seems to be
given by those forbidding facts that the proportion of blacks as
a total of the suburban population actually *declined* from 1950
to 1970, while the percentage of whites increased greatly; and
the converse in central cities. Clearly, the separation into two
nations *is* proceeding apace.

But this is the large and general picture, and if it is used to
argue simply that blacks are being confined in central cities, it
is false. The matter is far more complicated, and it is not easy
to give a short answer. One problem is that the "suburbs" we
can identify from the census are remarkably various. They
include, in many areas, open and lightly settled agricultural
land, which rarely has had any blacks on it in the North and

West, but has been occupied by a rural black population in the South. They include old suburbs that may be industrial and working class, as well as the more typical newly developed residential areas of higher income. Many central cities, on the other hand, include large tracts that are lightly settled and are being developed in "suburban" fashion. The growth of the black population in such central cities is not necessarily a sign of increasing "ghettoization." Even when we have large concentrations of blacks in central cities, they may not be "ghettoized" but spread broadly through the city, while small populations of blacks may be sharply concentrated.

Perhaps the most important factual issue is whether the concentrations of blacks that exist are the product of economic incapacity—in which case we may expect, with the improvement in the black economic position, that concentration will decline as more and more residential areas become economically available; or whether, alternatively, they are the result of discrimination against blacks or the result of taste. We have some evidence on all these issues, but hardly enough.

To begin with the question of suburbanization itself: The fact that the proportion of blacks in the suburbs has maintained its (relatively) small share—we find only half the number of blacks in the suburbs we would expect on the basis of even distribution, and twice the number we would expect in the central cities on the basis of even distribution—can itself be considered a matter for some congratulation. Blacks have slipped as a proportion of suburban population in twenty years from 6.1 to 5.3 percent; but this occurred at a time when suburbanization of the metropolitan areas was proceeding at a very rapid rate. The *absolute* number of blacks living in suburbs increased rapidly—from 2.2 to 3.7 million—and became a larger proportion of the black population.

While the situation from metropolitan area to metropolitan area was remarkably diverse, in nine out of the twelve leading metropolitan areas, in terms of Negro population, the increase

137

of black population in the suburbs proceeded more rapidly, and in most cases much more rapidly, than in the central cities (see Table 2). In absolute terms, these increases were not small.

TABLE 2

Increase in Black Population, in Absolute Numbers and Percent, in Central Cities and Suburbs of Twelve Metropolitan Areas Containing Largest Numbers of Blacks, 1960–1970.

	Percentage increase in central city, 1960–1970	Percentage increase in suburbs, 1960–1970	Black suburban population	
			1960 (1,000s)	1970 (100s)
New York	53	55	140	270
Chicago	36	65.5	178	128
Philadelphia	23	34	142	191
Los Angeles	52	105	117	240
Detroit	37	26	77	97
Washington	31	98	84	166
Baltimore	29	16	60	70
Houston	47	8	62	67
St. Louis	19	54	81	125
Newark	50	64	86	141
Cleveland	15	453	8	45
San Francisco–Oakland	40	61	68	109

Source: U.S. Bureau of the Census, *Census of Population and Housing: 1970 General Demographic Trends for Metropolitan Areas, 1960 to 1970, Final Report PHC (2)–1 United States,* Washington, D.C.: U.S. Government Printing Office, 1971.

By now, there are substantial black suburban populations in most metropolitan areas, and one may expect the factors that have contributed to their growth in the past—the breaking down of discriminatory barriers, the attraction of better and cheaper housing in the suburbs, and the decline in central city housing and services—to contribute to the disproportionate increase of black suburban populations in the 1970s.

One may view the figures on a greater black growth in the suburbs of large metropolitan areas with skepticism: After all, is it not true that black expansion in the suburbs tends to be either the expansion of the black ghetto, hitting the edge of the city and expanding into neighboring suburbs, or the expansion of small black settlements previously established in industrial suburbs or in the low-rent area of affluent suburbs? This is in large measure true, but it is by no means the whole story of black expansion into the suburbs. Thus, the huge percentage increase in the black suburban population of Cleveland reflects the fact there were almost no suburban blacks in that metropolitan area in 1960, and a good part of that increase is concentrated in one community, East Cleveland, whose black population increased from almost nothing in 1960 to 23,000 in 1970—in the process, the percentage of blacks rose from 2 to 51. East Cleveland may be viewed as an extension of the black "ghetto," but even so, it is an extension into an area of better housing and it brings some of the advantages of suburbanization to its black population.

Perhaps the best analysis of the black condition in the suburbs is that of Bernard Frieden.[3] It reveals that blacks in suburbs outside the South (where, owing to the pattern in which the outskirts of the cities contained a large population of poor rural blacks, the "suburbs" are often nearby rural sections) do have some suburban character. They have better occupations, higher incomes, more home ownership, and a lower proportion of female-headed families than central city blacks. Black suburbanites are not as well off on these measures as white suburbanites, but they are also not simply central city blacks moving over the city line. Suburbanization of blacks is what we commonly think suburbanization to be, and is not the extension of the ghetto.[4]

Of course, the gross analysis of trends obscures the very important differences between regions, and between different metropolitan areas. Suburbanization of blacks in one metropolitan area may be quite different from suburbanization in another. In one case, it may well be simply the ghetto spilling over. In Washington, to take one marked example, suburbanization for blacks means the movement of middle-class blacks into areas that are as attractive to them as they are to middle-class whites. An analysis of the 1960–1970 census points out that, for the first time in this century, the percentage of the black population in the suburbs of Washington rose, that the black population rose at a faster rate in the suburbs than in the city, and that the rate of black population growth in the suburbs was higher than that of the total population in the suburbs. In Prince George's County, blacks formed 14 percent of the county's population. In this county, blacks and whites are not far from each other in income (over $15,000 a year—blacks 23 percent, whites 36 percent; under $4,000 a year—blacks 9 percent, whites 6 percent) and in education (proportion of males graduated from college—blacks 11.9 percent, whites 12.6 percent). It is interesting that a massive desegregation by busing was ordered in this county not long ago. Whether in a county in which blacks and whites are so close economically this is necessary either for integration or for any educational effect is an interesting question. If it does have an educational effect, nothing in our social research today gives us any clear idea as to why it should happen.[5]

There are similar developments in other suburban areas and in newer areas of central cities. Scattered evidence in New York City and in its suburbs suggests that substantial integration is under way there. Thus, a large development, Starrett City, which will house 25,000 people, is going up in Canarsie, a section of Brooklyn in New York. According to the manager of Starrett City, quoted in the *New York Times,* "Based on the applications so far, my opinion is the development will be 70% white and 30% minorities." [6]

A fierce struggle was waged over busing black children into Canarsie schools from a distant low-income housing project in 1972. A report on that struggle in the *Race Relations Reporter* begins, "Canarsie could be said to have been built on racial fear." [7] But the biggest project going up in Canarsie will be 30 percent minority. One wonders, in view of the changing racial composition of Canarsie, whether the struggle of 1972–1973 was really worth it, just as one wonders, in view of the rapid residential integration of Prince George's County, whether the massive desegregation by busing of that county was worth it.

To consider some of the specifics of Table 2: There has been no increase in the black suburban population in the Houston metropolitan area. This may be explained by the fact that Houston still has much room for expansion and thus provides many of the benefits of the suburbs within the central city. Thus the *white* population of Houston increased 25 percent between 1960 and 1970, an increase unique among major central cities, most of which showed heavy declines of white population. Houston as a central city is still rapidly growing; that the black suburban population has not increased there may mean that blacks find opportunities for housing, services, and work in the Houston central city that are elsewhere to be found only in suburbs.

The black population of Detroit and Baltimore shows, against the general pattern, a greater growth in the central cities than in the suburbs. This may be explained by more severe resistance in the suburbs of those areas to the entry of blacks, but it is also true that great quantities of housing are available in those central cities because of the rapid decline in the white population. The two factors may, of course, be related.

The large increases in the number of blacks in central cities should not be taken as evidence that their economic circumstances are degraded or that they remain segregated. The economic rise and the desegregation of the black population is not to be revealed only by the progress of suburbanization. Inevitably—and this bedevils the entire discussion of this issue— we confuse race and class. Even if blacks are, on the average, poorer, and the poor are disproportionately black, the two cannot be used as substitutes for one another. That central cities are becoming increasingly black does not necessarily mean they are fated to become increasingly poor. A recent analysis of New York by the First National City Bank argues, "The city has not lost its middle class, as so many people suppose. It is true that many middle-class families move to the suburbs, often to be replaced by lower-income newcomers, but many other families moving up to middle-income status remain in the city. The income figures show only a germ of truth in the popular mythology." [8] The point is that even as the cities become more black, they are not becoming more lower class to the same degree, because the class composition of blacks is changing rapidly. More black need not mean more poor, and it means it considerably less today than in the past.

Nor does more black mean more segregated, a common illusion. Perhaps the most important work on the degree of segregation of blacks (as compared to other groups) is that of Karl and Alma Taeuber. Starting with *Negroes in Cities* [9] and continuing through many articles, including most recently unpublished analyses of the 1970 census, they have computed segregation indices for metropolitan areas by block. The segregation index tells us how far from even the distribution of a group is. If it is perfectly evenly distributed, the index will read zero. If every block is either all Negro or all white, the index will read 100.[10] For 1940 to 1970, the census permits the calculation of segregation indices only between whites and nonwhites; for 1970, one can also calculate them between white and black. The segregation of non-whites, the Taeubers had demonstrated, was unique in that it was extremely high, compared to other ethnic groups, and had remained high or risen while that of immigrant groups had declined. Certainly greater discrimination explained this difference; but it could also be understood as the result of the fact that between 1940 and 1960, when their earlier measures were computed, black city populations were *rising* while white ethnic immigrant populations were *falling*. One would expect, all other things being equal, a falling population to become less concentrated residentially, a rising one to become more concentrated. While no great comfort can be taken from the new figures for 1970—they continue to show on the whole a very high measure of segregation of non-whites, and only slight declines since 1960 —there is a strong reversal of trend. Despite the fact that the black population in all these cities has, one can assume, increased, the measure of segregation of non-whites has *declined*

TABLE 3

Indices of Residential Segregation
between Whites and Non-whites,
1950–1970, for Central Cities
and Selected Suburbs.

	1970	1960	1950
Boston	79.9	83.9	86.5
Cambridge	52.6	65.5	75.6
Chicago	88.8	92.6	92.1
East Chicago	79.0	82.8	79.6
Evanston	78.3	87.2	92.1
Los Angeles	78.4	81.8	84.6
Pasadena	75.0	83.4	85.9
New York	73.0	79.3	87.3
Mt. Vernon	78.4	73.2	78.0
New Rochelle	70.7	79.5	78.9
Yonkers	68.0	78.1	81.7
Newark	74.9	71.6	76.9
East Orange	60.8	71.2	83.7
San Francisco	55.5	69.3	79.8
Oakland	63.4	73.1	81.2
Berkeley	62.9	69.4	80.3

Source: Annemette Sørensen, Karl E. Taeuber, and Leslie J. Hollingsworth, Jr., *Indexes of Racial Residential Segregation for 109 Cities in the United States, 1940 to 1970,* Institute for Research on Poverty Discussion Papers, University of Wisconsin at Madison, Studies in Racial Segregation, No. 1.

in all but six of 109 cities. Between 1950 and 1960, on the other hand, the measure of segregation *rose* in forty-five of these 109 cities.

We do not have segregation measures for suburbs, for block statistics are not reported for many of them. Smaller cities on the outskirts of larger cities—generally "older" suburbs, where the black population has begun its entry to the suburbs—generally show *lower* measures of segregation than central cities, and more rapid declines in measures of segregation than central cities. What indeed these scattered figures suggest is that some suburbs are not more "segregated" than central cities (see Table 3).

The analysis of segregation indices is an arcane branch of the social sciences, and one hesitates to make much use of them. There is other evidence however which suggests that the popular image of a residentially segregated black population with no contact with whites is rather exaggerated. Thus, a 1965 survey reports the following experience of residential mixing within blocks:

TABLE 4

Percentage of Whites with Experience on Interracial Blocks

| | Region and Education (Percent) | | | | | |
| | South | | | North | | |
Negro family lived on same block:	Grade school	High school	Col-lege	Grade school	High school	Col-lege
Yes, live there now	6	3	5	15	10	15
Yes, used to	21	27	14	24	25	22
No, never	73	70	81	61	65	63

Source: M. A. Schwartz, *Trends in White Attitudes toward Negroes,* National Opinion Research Center, Chicago, 1967, as given in Thomas F. Pettigrew, "Attitudes on Race and Housing: A Social-Psychological View," in *Segregation in Residential Areas,* Amos H. Hawley and Vincent P. Rock, eds., Washington, D.C.: National Academy of Sciences, 1973, p. 32.

Apparently, outside of the South, more than one-third of the American population either live on a block with Negroes, or once did.

Another NORC survey tried to find out the extent of integrated neighborhoods in the United States. In view of the widespread belief that an integrated neighborhood is one that exists briefly between the entry of the first Negro and the departure of the last white, efforts were made to exclude changing or "transitional" neighborhoods: Informants were asked to report whether their neighborhoods were ones into which they expected both whites and blacks to move during the next five years. The process of determining what is a neighborhood and an integrated neighborhood is, of course, very complex and the procedures of the social scientists who conducted this study will not be reported here. However, it is revealing of common perceptions that the authors report ". . . we found that [city-wide] leaders consistently underestimated the number of integrated neighborhoods. . . . They classified as Negro segregated [changing (that is, to Negro)] some neighborhoods that local leaders and residents considered integrated because both Negroes and whites could move and were moving into them." [11]

They report: "On the basis of our data, we estimate that 36 million Americans, or 19 percent of the population, lived in racially integrated neighborhoods in 1967." [12] In view of the definition of "integrated" used, most of these neighborhoods contained very few Negroes: Only 7 percent of the Negro population was estimated to live in such neighborhoods. These figures must have risen substantially since 1967.

The University of Michigan Institute for Social Research asked national samples in 1964, 1968, and 1970 what the racial composition of their neighborhoods were. The number of whites answering "all white" dropped from 80 to 73 percent in these six years; the number of blacks answering "all black" dropped from 33 to 22 percent. These samples were asked about the racial composition of the school nearest them: whites saying (of grade schools) "all white" fell from 59 to 36 percent, blacks saying "all black" fell from 40 to 13 percent. The answers for high schools fell from 43 to 22 percent for whites ("all white"), and from 36 to 8 percent for blacks ("all black").[13]

Undoubtedly, in the five years since the last study reported, residential integration has proceeded further. The *perception* that one's neighborhood school was segregated seems much less extensive than segregation in schools as it is defined by civil rights agencies and litigants for desegregation.

Our conclusion from these facts is that when we speak of black residential segregation, we are talking about a problem that undoubtedly exists, but that is also much less of a problem than is generally assumed.

III

Effects: The effects of a possible redistribution of blacks more closely to approximate an even distribution between suburbs and central cities can be various indeed.

One expected effect would be an equalization of educational achievement through access to the more expensive schools to be found in many of the suburbs. As we know, the question of whether and when better education might be expected to follow as a result of a more even racial balance of children in the schools finds no simple answer (see pp. 120–123, Chapter 3). While suburbs do not necessarily spend more on schools than central cities, many suburbs of some of the largest cities with the heaviest concentrations of blacks do. If more blacks lived in these suburbs, more of their children would benefit from higher expenditures on them for education. This access to more expensive education might also come about from a successful legal assault on unequal expenditures in different school districts. This legal assault has failed for the moment in the Federal courts, but it may succeed in various state courts. However, the relationship between expenditure and educational achievement remains unclear. In the most striking case, New York State spends 161 percent of the national median per child, while California spends near it; yet, New York gets no more in educational achievement for the greater expenditure. Any expectation of a rapid rise in educational achievement for blacks owing to a massive shift to suburban schools is not solidly based.

A second effect that may be traced out is the impact of redistribution on black political power. This is now rapidly rising, but is by no measure near the 11 percent of the population blacks form. Would a redistribution increase or reduce black political power? Here the answer would appear to be unambiguous: Clearly, it should reduce it. Overwhelmingly, black political representation is based on numbers concentrated in constituencies. While we have the example of a Senator Brooke and a Mayor Bradley elected from constituencies with small minorities of black voters, most black representation is from districts with a heavy concentration of blacks. Undoubtedly, there would be offsetting gains. More blacks in the suburbs would mean that more political representatives would have to consider the significance of black votes and dominant black opinion, but in this area of possible effect, one would have to record a minus for blacks.

In yet a third area, I believe one should record a plus: better social relations and whatever follows from it; if, that is, the redistribution is of working-class or middle-class blacks into areas of similar status. We have much evidence of good social relations between blacks and whites in interracial neighborhoods. By their nature, social relations cannot be easily summarized or reduced to figures. One assumes good social rela-

147

tions may mean access to important networks of information (as to jobs, housing, education), better race relations generally and a reduction of interracial conflict, and, aside from any instrumental gains, they may be rewarding in their own right. Yet as we know, there is great resistance in many neighborhoods to the entry of individual blacks of the same social and economic level, and enormous resistance to blacks of a lower social and occupational level. The management of integration to achieve the positive results and avoid the negative ones is no simple matter. Adding to the difficulty is the fact that the present offensive to open up the suburbs does not concentrate on getting the middle-income black to be able to join his economic equals in the suburbs: The lawyers working on these cases assume he can already get there on his own, even with some effort (in any case, what he needs to get there is stringent enforcement of the laws against housing discrimination). The present conflict is over the entry of low-income people into the suburbs, for it is they whose entry is limited by local jurisdictions' zoning regulations. It can be confidently predicted that success in *this* endeavor will do little to improve race relations.

One improvement to be expected from a black move into the suburbs is better housing. Thus, there is a much greater rate of homeownership among blacks in the suburbs. On the other hand, there is, surprisingly enough, more substandard black housing in the suburbs than in central cities. This is undoubtedly owing to the fact that for blacks, as for whites, "suburbia" as the census defines it may mean very different things, from depressed rural slums to affluent middle-class housing.[14] Yet, if one image which follows from thinking of blacks penned up in the central cities is that it must be very difficult for blacks to improve their housing, it is wrong. The fact is that the reduction of "substandard" housing between 1960 and 1970 among blacks ("substandard" is defined as a combination of housing lacking complete private plumbing facilities, plus housing that is considered "dilapidated") was at the same rate as among whites. In both cases, the reduction was nearly 50 percent—in 1960, one-half of the housing occupied by blacks was substandard; in 1970, one-fourth.[15]

Perhaps the most important effect hoped for from an opening up of the suburbs is more rapid improvement of the economic position of blacks. This would follow from the well-known analysis of the dynamics of the distribution of jobs in metropolitan areas, which argues that job growth is concentrated in the suburbs and increasingly so, that blacks concentrated in central cities without good transportation facilities leading to these areas of job creation are cut off from access to these jobs, and that the jobs that are left in the city are increasingly those of higher skill (it is the manufacturing jobs, in their search for large expanses of open land on which to build, and the service and sales jobs, following the movement of population, that head for the suburbs), which are cut off from minorities owing to their poorer education and skills—the job "mismatch" hypothesis. It is less well-known that, under recent reexamination, this entire analysis has been seriously weakened. Thus, Bennett Harrison, in *Urban Economic Development*,[16] analyzes a substantial new body of data as well as the older data which led to the thesis I have just summarized and argues that (1) there is no evidence that there is any acceleration in the process by which jobs (and population, tending to move together with jobs) grow more rapidly in the suburbs; (2) that the number of jobs in relation to labor force has, if anything, increased in the central cities in recent years; (3) that there is no evidence of a job mismatch.[17]

Naturally, reversal of such widely accepted axioms will be treated with skepticism, but the fuller argument makes sense: The data on which the early analysis was based was postwar, and ended with figures from around 1963. In the early postwar period, there was a particularly rapid decentralization of jobs, owing to the resumption of long-term trends interrupted by depression or war. The late 1950s was a period of slow economic growth, in which (according to Benjamin Cohen), because of high excess capacity, firms close down their least productive plants which are in central cities, and locate their new plants in areas which would be most attractive to professional and managerial employees—there is no problem in such periods in getting less skilled workers. Thus, interestingly enough, analyses based on data up to 1963, becoming current in the latter 1960s, were challenged by developments after 1963.

On the other hand, there is the decisive evidence from the census, given by Downs, for 1960–1970, that in the central cities of the fifteen largest metropolitan areas, civilian employment *did* decline by 836,000. But a lot depends on which years and metropolitan areas one selects, and the massive growth of *government* jobs changes the picture: "In the thirty largest metropolitan areas total local government employment increased enough between 1957 and 1962 to more than offset job losses in other sectors. After 1962, the decline in private employment was reversed in the central cities of most of the large metropolitan areas: in the thirty largest areas, taken together, private employment rose by an average of 99,000 jobs per year between 1963 and 1967, compared with an average decline of 6,000 between 1958 and 1962." [18] Analyses of the effects of post-1969 recessions on the relative growth of jobs as between central cities and suburbs may again show a sharper shift of jobs to the suburbs, but the fact is that the story has not been all one way, and even more striking, the theory of job mismatch, reasonable as it appears (and true as it may be, for example, for New York), does not survive a general analysis of job patterns in metropolitan areas.

In any case, as many point out, the jobs are still basically located in the central cities—which is why the dominant commuting trends are into them. This has not apparently affected the pattern of higher black unemployment. Harrison seems to feel this is owing to discrimination; Richard Freeman be-

lieves discrimination is no longer a factor and it is due basically
to poor endowment—education and family background.[19]

I V

Policies: Our discussion of policies to open up the suburbs
must, of course, be influenced by our analysis in previous pages
of the actual patterns of suburbanization of minorities and the
actual effects to be expected from a higher rate of suburbanization induced by public policy.

Once again, we must insist on the reminder: The enforcement of the laws against discrimination in housing is not in
question. Indeed, in this area a more intensive and sustained
effort to wipe out discrimination seems in order, for discrimination in housing appears to be much wider and more serious
than discrimination in education or in employment. Complaints
to state agencies and the Department of Housing and Urban
Development are processed slowly, and the individual seeking
housing is often reduced to frustration and bitterness.

Under Title VIII of the Fair Housing Act of 1968, all housing except that sold by an individual without the assistance of
a broker or advertising or that rented by an individual in a
house in which he lives (and which does not contain more than
four units) must be made available without discrimination on
account of race, color, religion, or national origin. During the
first nine months of fiscal year 1973, HUD was receiving
about 230 complaints a month. It can only deal with them
through investigation and conciliation, and this takes time.
(There are also many state and city agencies that handle housing discrimination complaints, some with greater powers than
HUD.) Under Title VI of the Civil Rights Act of 1964, no
government benefit may be provided on a discriminatory basis,

and this applies to the benefits of HUD—housing subsidies, urban renewal and community development funds, sewer and water grants, planning assistance, mortgage insurance, and all the other benefits to assist planning, urban development, and the provision of housing. At the beginning of 1973, HUD had about 200 complaints on hand. It has the power to cut off funds, but this is rare and HUD prefers to negotiate an agreement that will bring a local agency into compliance with Title VI. Under Title VIII of the Fair Housing Act of 1968, the Department of Justice can bring suit against anyone engaged in a pattern or practice of discrimination. Fifty-eight such suits were filed during fiscal year 1973. The central and regional Equal Employment Opportunity offices of HUD contained 427 positions, and HUD officials said this was insufficient. Aside from these, there was a legal staff in the Department of Justice engaged primarily in housing and community discrimination cases.[20]

It is clear, however, that enforcement of laws against discrimination, no matter how stringent, will have only a modest effect on residential distribution. Those concerned with a more even distribution of minority groups throughout the metropolitan area must resort to stronger measures, and the battle over the opening up of the suburbs concentrates on these stronger measures and their legitimacy. As in the case of employment and education, the question arises: Is an uneven and unrepresentative distribution itself to be taken as a sign of discrimination? Can government agencies and courts act against an uneven distribution, taking it as evidence that discrimination by someone—builders, sellers, real estate agents, landlords, banks, government agencies, local governmental authorities, and so on—has been responsible for creating it, and that real action against discrimination necessitates achieving equal representation or even distribution? We will discuss shortly the development of government regulations and law in this area, but let us first be clear about what may cause uneven distribution of minority groups in metropolitan areas.

There are at least three quite different factors which lead to residential concentrations of ethnic groups: one is discrimination; a second is economic—the restrictions imposed by limited income, and the choices favored owing to the jobs in which the group is concentrated; the third is cultural—the satisfactions found in a community, which requires some degree of physical closeness to other members of a group and some degree of concentration to support community institutions. In the case of employment and education, as we have seen, judges, lawyers, and government officials operate on the assumption that if there were no discrimination, an even distribution of blacks or any other ethnic group in the occupations and educational institutions of a society would occur, and so variation from even distribution is suspect as a sign of discrimination. We find the same assumptions in much of the discussion of ethnic residential concentrations, and following from this error, an attempt to institute even distribution as a means of overcoming a presumed discrimination.

It is not easy to give any statistical demonstration of what part of black residential concentration is due to discrimination. There have been efforts to determine the part of residential concentration that may be explained by economic factors, and estimates vary considerably.[21] Whatever the conclusion as to the weight of economic factors alone, however, it is impossible to separate from the remaining part of the concentration to be explained the part owing to discrimination and the part owing to culture: taste, wishes, preferences. There are those who argue (as does Judge Roth in Detroit—see Chapter 3, p. 106) that no part may be owing to culture—that in the United States any group should distribute itself perfectly evenly in the absence of economic and discriminatory barriers—but the fact is that groups that have suffered much more modest discrimination than blacks, and some that have suffered no discrimination at all, are also concentrated.

Nathan Kantrowitz has made a careful effort to determine,

as best one can, whether concentration of blacks in 1960 in the New York metropolitan area could be considered similar to that of other ethnic groups. One serious problem here is while all blacks are recorded by the census, in the case of European ethnic groups only foreign-born and children of foreign-born ("foreign white stock") are recorded. The question he asked is not how are blacks distributed in relation to *whites* but how are blacks distributed in relation to the various ethnic groups that make up whites, and how are these distributed in relation to each other? Kantrowitz considers as an upper limit of what we might expect in the absence of discrimination the degree to which the residential distribution of the more recent immigrant groups (from Eastern and Southern Europe and their children) differentiates them from earlier immigrant groups (from Northwestern Europe and their children). The average "segregation index" summing up the differences in residential distributions between Northern and Southern European groups is 51.6. "It seems reasonable," he writes, "to consider this average index of 51.6 as the lower [possible] bound for 1960 segregation of foreign white stock from blacks or Puerto Ricans."

In effect, nearly 40 years after the end of large-scale European migration, a segregation index number encompassing both the immigrants and their children (but primarily the children) in a highly suburbanized metropolis indicates that, on the average, 51.6 percent of the population of Southern European origin would have to be redistributed in order to achieve full integration with the Northern European population. . . .

In order to indicate what this statistic means, he writes:

Suppose blacks were white—that is, imagine a post-World War II mass immigration to New York of a Caucasian population unique in some culture traits, such as religion or language, burdened by backgrounds of rural or small-town poverty and lack of education. . . . One can hardly expect the segregation of these newcomers to be lower than the 51.6 that has just been seen between Northern and Southern Europeans. By simply adding to this lower bound some margin to account for the poverty and recency of migration of these newcomers, a segregation level of at least 60 would probably be reached.

The actual segregation index figure for the separation of blacks from white immigrant groups, new and old, in 1960 was 81. Puerto Ricans showed the same average separation from white immigrant groups—80.2. Blacks were almost as segregated from Puerto Ricans as each group was from other white groups —66.0.[22] This analysis suggests that a realistic figure for black concentration in the New York metropolitan area in 1960 in the absence of discrimination would have been 20 points lower. (As we saw above, the segregation index for blacks declined in the New York metropolitan area between 1960 and 1970.)

To attempt to eliminate through public policy all concentrations of blacks and other minority groups would clearly be to attempt to undo far more than discrimination alone has created. Of course, as we have seen, there are other objectives of an even distribution of blacks and other minority groups aside from that of simply undoing discrimination—better race relations, access to jobs, and educational opportunities—and it is worth pursuing these objectives, but one cannot do so on the ground one is undoing the effects of discrimination.

Discussions of policy should not only consider what should be done—whether from a moral or a pragmatic perspective—but what can be done. Many working toward more even distributions are unaware of how difficult an objective this is in a society in which almost one-fifth of all families move every year and one-half move in any five-year period. What these figures mean is that neighborhoods can change remarkably rapidly. Thus any stable situation in terms of some distribution of racial and ethnic groups is not easy to attain. Overcoming discrimination, in view of the mobility of the American population, will not necessarily mean stable and even distributions. Thomas Schelling, in a brilliant article, has demonstrated how difficult it is in a mobile situation to achieve any given result. Take, he says,

. . . a roll of pennies, a roll of dimes, a ruled sheet of paper divided into one-inch squares, . . . and find some device for selecting squares at random. We place dimes and pennies on some of the squares, and suppose them to represent the members of two homogeneous groups—men and women, blacks and whites, French-speaking and English-speaking. . . . We can spread them at random or put them in contrived patterns. We can use equal numbers of dimes and pennies or let one be a minority. And we stipulate various rules for individual decision.

For example, we could postulate that every dime wants at least half its neighbors to be dimes, every penny wants a third of its neighbors to be pennies, and any dime or penny whose immediate neighborhood does not meet these conditions gets up and moves. Then by inspection we locate the ones that are due to move, move them, keep moving them if necessary, and when everybody on the board has settled down, look to see what pattern has emerged.

This exercise, it turns out, leads to "segregation"—all the pennies in one area of the board, all the dimes in another. We can change the rules, the relative proportions: The results are the same.[23] If groups are differentiable, and if they have even modest tastes affecting their behavior, a stable and even distribution is hard to achieve. Even without these ingenious exercises, one may demonstrate the difficulty. In the study by Bradburn (see p. 145), neighborhoods that were "stable" in the sense that both whites and blacks were expected to be willing to move into them for at least the next five years were considered integrated. Seven percent of all blacks, 19 percent of all whites, lived in such neighborhoods. Only 3 percent of white households, it was estimated, lived in neighborhoods with more than 10 percent of Negroes. A good number of whites may consider 10 percent black too integrated; a good number of blacks may consider 10 percent black as not quite integrated enough. One is reminded of a study of a Jewish community about twenty years ago in which the most common response as to a desirable religious distribution in a neighborhood was about 50 percent Jewish; but this is not the answer most non-Jews would then have given, and so neighborhoods, once they started being comfortable for Jews, started being uncomfortable for non-Jews. The situation may be similar with blacks. A 1968 survey of blacks reports that 48 percent would prefer a neighborhood mixed half and half, 13 percent all or mostly Negro, 1 percent mostly white, and 37 percent no difference.[24] The dynamics of black residential behavior may be such that the achievement of, let us say, a neighborhood that is "only" 10 or 20 percent black risks instability and a movement to mostly or all black.

Thus some kind of *conscious* management of movement by race is necessary to increase the number of stable integrated neighborhoods. There are examples of this type of management, of affirmative action for integration. If managers see a development or project becoming heavily black or heavily white on the basis of free choice, they can energetically recruit members of the group needed for a better racial distribution. There are limits to what may be done in this way. The "benign quota" (the setting of numbers for one group or another) is unconstitutional; and further, the effort to achieve racial balance by favoring whites over blacks, which may be necessary to achieve a *stable* distribution, is indistinguishable—if someone is willing to take the matter to HUD, or the Justice Department, or court—from simple discrimination, which is also illegal. Nevertheless, subtle measures in this direction are possible and through them a higher measure of integration is possible.

There is a final policy problem that we will find complicating our understanding of developments in this area, and that is the relationship between policies designed to produce an even distribution by race and the resulting distribution by income. Discrimination on grounds of race, color, religion, and national origin is illegal. Discrimination on grounds of income, wealth, or credit is not. In the fight to open up the suburbs by means of a breakthrough in constitutional law or Federal regulations, the effort is to demonstrate that discrimination on grounds of economic incapacity is equivalent to discrimination on grounds of race, for the latter class is protected by the Constitution and the laws, the former is not. If one makes a breakthrough for the black, one has made a breakthrough for the poor. The benefit of residential integration further is seen as necessary for the poor—whether black or white—as well as for the black. There is a double objective here, and progress in integrating one category is seen as leading to progress in integrating the other. And yet, of course, the two categories are not identical. The middle-class black will resist the entry of a low-income housing project just as the middle-class white will, even if the housing project will be black. Indeed, the key legal decision limiting the Federal government's power to assist in the building of low-cost housing projects in an area in which a concentration of such housing threatens a middle-class area was brought by blacks (*Shannon* v. *HUD,* 436 F. 2d. 809 [3d Cir. 1970]). This is perhaps the strongest mandate now in existence limiting concentrations of low-income groups and requiring some broader distribution of these groups through the larger urban area.

In the area of housing and residential discrimination, hard affirmative action has not proceeded as far as it has in employment and education. By its nature, one would think, it could not. People are not assigned to housing by central bureaucratic institutions, as they are assigned to schools; nor do large central organizations equivalent to large employers exist in housing or residential distribution with the power to alter directly ethnic proportions, as employers may. Housing is still for the most part a matter of free choice limited by economic capacity and tastes. But the progress toward hard affirmative action proceeds in this area, too, pursued through pressures on the Federal government, the states, and local jurisdictions, primarily through court cases. A mild form of affirmative action is already well established: It requires large builders and developers with government assistance (subsidies or mortgage guarantees) to develop an affirmative action marketing plan. In a situation where so many in minority groups fear that they are not welcome, this is valuable and indeed essential. It is equivalent to the clear assertion of an employer that equal opportunity prevails and that every applicant is welcome. Even this, however, can be taken as an opening wedge to quite a different form of affirmative action. For example, the Civil Rights Commission, following its position on affirmative action in employment, asserts ". . . the most important determination to be made through monitoring [affirmative action plans] is the extent to which anticipated results have been met. No matter how much advertising has taken place, if racial and ethnic minorities are not purchasing homes in the subdivision, the plan being reviewed is not successful and the marketing and sales techniques being used will warrant careful scrutiny." [25] One sees the process whereby affirmative action became goals and targets, and goals and targets became quotas, in its incipient stages in this comment: The notion that blacks or Mexican-Americans or Puerto Ricans or Indians may simply prefer to live elsewhere no matter how persuasive the sales techniques does not seem to occur to the Civil Rights Commission.

There are limits, however, to what the Federal government can do in this area.

Our early housing policies gave powers to local government to accept or refuse subsidized public housing. Large cities accepted it; smaller cities often did not or, if they did, accepted it only for the aged. Furthermore, in many communities there are referenda as to whether a given project should be undertaken. Such referenda were attacked as racially discriminatory in themselves in *James* v. *Valtierra,* but the Supreme Court refused to overrule such local powers (402 U.S. 137 [1971]). In 1968, with the 235 and 236 mortgage-subsidy programs, local public agencies confined to city boundaries were by-passed. Any builder, non-profit or profit, could build such a project. This new approach was undertaken for a number of reasons, among them dissatisfaction with the public-housing projects managed by local public agencies and the belief that the poor should have access to home ownership. These programs were enormously successful and produced far more housing annually than was ever built under public housing. But there was also room for scandal, they were quite expensive, and in January, 1973, they were suspended by the Federal government. They have been replaced in the Housing Act of 1974 with a new approach to low-cost housing, one which permits public housing authorities as well as profit and non-profit builders to build housing for the poor under a leased-housing approach. With the shift to the 235 and 236 programs, the Federal government had to confront the problem of what to do about communities that excluded 235 and 236 on the basis of zoning and other powers, whether to keep out blacks or the poor. On the whole, it did little, and there was little it could do if it wanted to see housing built.

President Nixon, in 1971, asserted that the antidiscrimination laws would be enforced but the administration ". . . will not attempt to impose Federally assisted housing upon any community." In 1972, HUD developed "project selection criteria" to guide HUD in approving subsidized housing projects proposed under all programs. As is common, these were initially presented for public comment, and as the department reported in presenting its criteria, ". . . some comments asserted that the project selection criteria will result in too few projects being built in the inner cities, and others asserted that the criteria will result in too many projects being built there." The criteria (37 C. F. R. 203, "Project Selection Criteria") would strike any objective reader as a demonstration of the absurdity of government trying to solve all social problems, however defined, simultaneously. The first criterion is easy—whether there is a need for low income housing in the area. The second—". . . to provide minority families with opportunities for housing in a wide range of locations"—does not really settle the question of whether the department should favor projects in areas with few minority families (to assist integration) or in areas with a good deal (to permit minority families to get housing where they lived). The third criterion, titled "improved location for low[er] income families" is ". . . to avoid concentrating subsidized housing in any one section of a metropolitan area or town." Whether operationally doing so does offer "improved location for lower income families" could occupy an army of sociologists. And so it goes. Criterion six, ". . . to produce housing promptly and to provide quality housing at a reasonable cost, . . ." seems to be in contradiction to numbers two and three. These project selection criteria will now apply to the new leased-housing program that the Housing Act of 1974 has established as the chief means of Federal assistance for low-cost housing.

Much of the responsibility for opening up the suburbs has, with the shift of more and more Federal housing and community development money into state revenue-sharing and the rise of state housing corporations, passed to the state level. Here the Urban Development Corporation of New York was a pioneer. Initially, it had the power to build anywhere and override local zoning controls, if it felt it necessary. But when it tried to do so, it raised such a storm that these powers were taken from it by the state legislature. Another approach, much studied, is that of Massachusetts, which has passed a law permitting appeals from local zoning to a state board, which can override the local decisions if they are taken in a community which does not have a certain amount of low-income housing. The efforts to make use of this mechanism have not been very effective up to now because even if a zoning decision is overridden there are other legitimate grounds for localities to exclude low-income housing. The new concern for environment and requirements for environmental impact reviews have given many local communities substitutes for zoning. Often low-income housing for the suburbs is proposed on cheap land, and it is just this land which may be swampy and provide nesting for birds, and so forth. According to a summary on May 19, 1974 (*Boston Globe,* "Opposition from Suburbia Stifles Antisnob Zoning"), the state Housing Appeals Committee had heard seventeen appeals, and nine more were pending. It had approved, subject to meeting certain conditions, eleven applications, rejected one, and was still considering the rest. Both the New York and the Massachusetts approaches (and others) have been proposed for other states, but political resistance has been severe and it is not clear how much more progress can be made along this route.

A new level of quasi-government now exists—the councils of governments (COGs) that have been set up under Federal prodding in metropolitan areas and that are required to review all applications from local governments in their area for Federal support. One of these, the Miami Valley Regional Planning Commission (covering the Dayton, Ohio, metropolitan area) pioneered in developing a regional housing plan in 1970, under which each local area agreed to take a certain amount of subsidized housing. This seems to have been a success in voluntary action, and the building of much subsidized housing outside of Dayton followed. Other COGs are trying to follow in the footsteps of the Miami Valley Regional Planning Commission.[26]

The key battleground, however, is the local municipality with its powers to zone and to set other standards that will limit the kind of housing that may be built in it and, consequently, the kind of people that may live in it. From the point of view of residents in self-governing communities, it is better to have rich people who pay high taxes than poor people who pay low or no taxes, better to have fewer families with few children to educate in expensive schools than to have many families with many children to educate. Alongside financial considerations leading to zoning restrictions are environmental ones. People who have moved to small, suburban communities have exchanged city densities for a great degree of openness and want to preserve low density. Thus, the popularity of large-lot zoning. And, of course, there are racial considerations. The nub of the matter is the extent to which these various considerations dominate or affect local zoning restrictions.

The assault on zoning and related local powers to restrict building is now going on in many courts. There have been successes in overturning one kind of restriction or another; there have been failures. But to date there is no *Brown*-type breakthrough which has simply put aside the power of local communities to zone and otherwise restrict development. What seems to have been determinative in preventing such a breakthrough to date is that, while racial groups are constitutionally protected categories and cannot be excluded, income groups are not and can. Issues of "intent" will be important here, and courts will differ. The greatest success to date in opening up the suburbs through this path is a New Jersey Supreme Court decision of March 24, 1975, which asserts: "We conclude that every municipality must, by its land use regulations, presumptively make realistically possible an appropriate variety and choice of housing." The decision was made under the New Jersey constitution and may turn out to be the breakthrough advocates of the striking down of local municipality powers are looking for.[27]

In this struggle, middle-class blacks have often fought along with middle-class whites against subsidized development for low-income groups. These new allies on the one side have been matched by new allies on the other: It is the builders and developers who are now allied with open-housing advocates

163

and public-advocacy lawyers in trying to overturn local powers to restrict development.

The most important case at the Federal level was that of the Federal government against the town of Black Jack (begun the same day President Nixon issued his policy statement against "forced housing"). The town of Black Jack sprang into existence—taking this most inappropriate name—in the metropolitan area of St. Louis because a subsidized, multifamily rental development was planned in the area. Black Jack rezoned the site for the development to prohibit multifamily uses. "There is a horrifying fascination," a planner writes,

. . . in reading the planning report and zoning ordinance prepared by the Black Jack Planning and Zoning Commission, a body without professional planners. . . . It is not surprising that citizens with no planning experience could draft a competent document. What is alarming is the ease with which the traditional planner's techniques and phraseology were used to support the decision to make Black Jack a single-family community. . . . The report relies on the "basic character of the community . . . being single family," and the "already overburdened traffic arteries" as justification for excluding [the development].[28]

The legal argument was that Black Jack was trying to exclude blacks in violation of the Fair Housing Act of 1968. To succeed, it had to be shown that Black Jack's actions excluded not only the low-income, but the black. Researchers of The Urban Institute demonstrated that the exclusion of the proposed development would have a disproportionate impact on blacks. Thus, while 57 percent of all blacks in the St. Louis metropolitan area lived in multifamily housing, only 25 percent of whites did, and while 15 percent of moderate-income black households lived in multifamily rental housing, only 8 percent of whites did.[29] This is the kind of statistical evidence that has led to judicial action in employment and education cases. On March 20, 1974, a Federal District Court refused to overturn Black Jack's decision to exclude multifamily housing. It said:

The evidence on this case does not support a charge of racial motivation, purpose or intent in the incorporation of Black Jack as a city or the passing of Black Jack's zoning ordinance. . . . In testing the validity of the action of the government of the city, county, state or country against a charge of racial discrimination, it would be ludicrous to contend that the feelings and personal prejudices of every resident or even governmental officer of that government must be examined. Personal feelings or prejudices unexpressed and unmanifested in action would become the test of the validity of governmental action. Although it is true very few officials will openly state racial reasons for their actions, racial considerations must be shown as part of the legislation or as a significant reason for the legislation. . . . Multiple reasons were advanced for opposition to apartments, among them the character of the community, road congestion, school impaction, property devaluation, and opposition to transient apartment dwellers. . . . These are valid state reasons to pass such an ordinance and supply a rational basis for the actions of the City of Black Jack.

This decision was overruled by the Eighth Circuit Court in December 1974. It found the statistics more persuasive.

Other approaches to opening up the suburbs, besides the favored legal route, are also possible. Anthony Downs proposes voluntary action stimulated by economic incentives—to low-income groups to move to the suburbs, to builders to build for them, to local governments to accept them. He quantifies the objectives of his proposal in terms of necessary migration flows, necessary amount of new building, and the like. And he nobly lists the range of strategies and tactics that would be necessary to implement a substantial level of movement of the poor into the suburbs, a listing that would lead most readers to conclude the task is impossible. Thus, the very *first* policy involves the strengthening of decision-making powers of old governmental bodies and new ones, and includes the creation of multifunction metropolitan development agencies, creating minimum sizes for governments exercising zoning powers, consolidating small suburban communities into larger ones, creating metropolitan agencies with stronger review, amendment, and veto powers over lower level decisions. There are nine other major policies, most with a substantial range of necessary tactics. Downs emphasizes economic rather than racial integration. He believes this may be easier to implement, and that black leaders now tend to oppose the dispersal of blacks.

But it is clear the subsidies necessary to get builders to build, low-income city dwellers to move, and suburbs to accept new low-income populations would have to be enormous. While we would all favor economic incentives rather than compulsion where possible to get people to act in the public interest, the history of the 235 and 236 programs—which would be only one small part of a Downsian strategy—suggests the costs would be overwhelming. Further, these measures would *increase* the rate of movement from the declining central cities, which now have excess housing, schools, hospitals, and transportation facilities. It is true Downs proposes simultaneously—as people are being encouraged to move out—improvement of inner-city areas (though with least emphasis on capital investment). But would that not only add to the cost?

Admittedly, it is all too easy to flinch in the face of necessary heroic measures. But are these measures necessary? I would argue they are not. The integration of blacks proceeds, and at a pace related to their rise in income and occupation level. The segregation of other minority groups is based more on income and occupation than on racial and ethnic discrimination and will decline with rising incomes and related changes in occupation and culture. The integration of the poor is quite another matter, and is hardly likely to be much advanced whatever measures of public policy we adopt. The poor are constrained in their movements by limited income, are resisted by the middle classes because of the social problems they bring, and further, it is not at all clear that the poor will be better off if distributed through an active public policy—even if it were possible—among the middle classes.

The problems raised by the debate on the opening up of the suburbs are similar to those raised by the debate over busing. In both cases, *voluntary* action may do a great deal in bringing us toward a more integrated society. But in both cases, voluntary action works for relatively stable families. It is they, one assumes, who are more willing to find new housing and neighborhoods in the suburbs, to take action to send their children to distant schools so they can improve their education. But in our cities there is another problem, and that is the large, depressed section of the black population. It is not identical with the huge female-headed family sector, or the welfare population, or the population out of which comes a disproportionate number of juvenile delinquents, the violent, and the disturbed. But the overlap among all these categories is substantial. The black middle class, the black worker, does not need any special help from government, aside from the strict enforcement of the laws against discrimination; these groups have made substantial progress and are steadily making more. There should be no problem for other Americans in fully accepting them as neighbors and accepting their children in schools; if there is, the full force of the law must be brought into play to replace prejudice and silliness with reason. But there is, alas, another problem, and all our experiments and new departures in social policy have made small headway, if any, against it. As long as this other problem exists, as long as it holds the key place in public consciousness that it does, as long as it continues to make the central city dangerous and unpleasant, and, to many, immoral to boot, I do not see how we can expect anything but resistance—from black and white—to the overruling of local governmental powers in order to open up the suburbs.

CHAPTER

5

The White Ethnic Political Reaction

I HAVE described the development of affirmative action into requirements for even statistical distribution in employment and education and the threat of such a development in housing. Many assert that this development is essential if the promise of real freedom and equality for the blacks and other American minority groups is to be fulfilled. I have argued in each case that equal opportunity, not even statistical distribution, is the proper objective of public policy. That argument can be made on constitutional grounds—and I have made it. It can be made on pragmatic grounds, for blacks made real and substantial and permanent advances while equal opportunity, supplemented by affirmative action for equal opportunity, was still the main thrust of public policy. And that argument can also be made on political grounds: that equal opportunity represents the broadest consensus possible in a multiethnic and yet highly integrated society, and that this consensus would be broken if requirements for statistical representation were to become a permanent part of American law and public policy.

One of the most visible signs of a strain to such a consensus has been the fate of the coalition of workers, white Southerners, Northern ethnic groups, and blacks that became, under Franklin D. Roosevelt, the majority Democratic Party of the 1930s and 1940s. The first major break in this coalition came in 1948 when the Democratic Party committed itself to civil rights for blacks, and a good part of the South broke away from the coalition. The second, however, came with the success of the drive to establish full equality for blacks: This second major break in the coalition, the "white ethnic backlash," is the subject of this chapter. It has suggested to many that racism was not a characteristic primarily of the American South, but of all America and all Americans, so that even later immigrant groups, infected by this pervasive racism, would fight equality for blacks. Indeed, many have argued, they oppose equality more fiercely than do old Americans, the white Anglo-Saxon Protestants. If this is so, we are faced with two unpleasant alternatives: Either we must undertake the same struggle against the racism of the white ethnic groups that was undertaken against the South, or we must accept a divided and unequal society. But before we consider these alternatives, it is necessary to examine the so-called white ethnic backlash.

The issue first arose sharply in 1964, when George Wallace, coming out of the South with his widely publicized resistance to school desegregation, began testing the waters in the Border States and the North. He found, to the surprise of many, a remarkable response beyond his natural ethnic base—white Southerners who had migrated to the North. In the Wisconsin, Indiana, and Maryland primaries of 1964, the Governor of Alabama scored 34, 30, and 43 percent of the vote, respectively, in the Democratic primaries.[1] Even before the first of the summer riots and the rise of the Black Power movement, Berkeley, California, had, by referendum, rescinded a fair-housing ordinance, and Seattle, Washington, had defeated an open-housing ordinance. It became more difficult to pass new fair-housing laws. In New York City, white parents organized against efforts to balance the schools racially. In Boston, Louise Day Hicks won election to the School Committee on the issue of resistance to desegregation.

The "white backlash" raised questions about the political attitudes of whites of recent European background, whites who had, under the Democratic coalition forged by Al Smith and Franklin D. Roosevelt, been allied with blacks and other minorities. Clearly, white Protestants of Southern and Northern background and the descendants of earlier Northeast European immigrants, were also involved in the white backlash. But what of the later immigrant groups and their descendants? David Danzig, analyzing Wallace primary victories of 1964 in Northern and border states, wrote:

. . . [Wallace] made a strong showing both in the ultraconservative suburbs . . . and in the rural areas which were former McCarthyite strongholds. But he scored almost as heavily in the Democratic wards of Milwaukee—the predominantly Polish, Italian, and Serbian neighborhoods that border on the Negro area of the city. Similarly, in the Indiana primary, Wallace carried Lake County, a heavily industrialized complex that includes Gary, Hammond, and East Chicago, and that has a large East-European and Negro population. In Maryland, Wallace won in the Irish and Italian districts of Baltimore. . . .[2]

That white Protestants should resist the advance of new groups, both in the North and the South, was nothing new. Nativism has a long history in American life.[3] That the resistance to black advances should be most strongly centered in the white South and white Southern migrants in the North was to be expected. But that ethnic groups themselves recently risen from poverty and powerlessness should be the center in the Northern and Western cities of the most intense resistance to further black advances: This appeared ironic and tragic.

And it was further a matter of enormous political significance. The white Protestant population of the North and West had been the chief source of support for the Republican Party. The white Protestants of the South, beginning in 1948, were engaged in a complex process of movement into the Republican Party, both directly and by way of intermediate rebellious Democratic groups. The loss of support among later white ethnic groups could presage a Republican dominance at least as overwhelming as the Democratic dominance of the Roosevelt years.[4] Those who proposed for the Democratic Party a strategy that would maintain dominance also saw the later white ethnic groups as crucial.[5] But the political behavior of these groups still remains something of a mystery, even though miles of computer print-out and reams of analysis have been devoted to it. The mystery is summed up in the question: Is there a distinctive reaction of members of white ethnic groups to the advance of blacks and other minorities? Is there a distinctive political interest that we may legitimately call white ethnic, so that it can be appealed to by Republicans, repelled by liberal Democrats, and called back with the proper appeals to a new Democratic coalition? If there is, what is it?

Our first problem is that we do not know very much about the "white ethnic groups." The census counts blacks, American Indians, Chinese, Japanese, and other races. (For one census, Mexicans were counted as a "race.") But ethnic origins are not, as such, recorded by the census. The country of origin of residents born abroad is recorded and so is that of the parents of all those who are children of the foreign born, which permitted a rough estimate of the size of an ethnic group as long as it consisted predominantly of immigrants and their children. These figures are increasingly less helpful as the age of mass immigration recedes ever further into the past. The United States Census did not attempt to gather any information on ethnicity, surprisingly enough, until 1969. The mere fact that the U.S. Census decided to collect such information—five years after national legislation prohibited discrimination on account of national origin—should be taken as one sign that an "ethnic revival" was going on, or that at least many people thought it was.

The census had to answer the question: What is an ethnic group? The answer to this question indicates how complicated the term "ethnic" is and how hard it is to present an unambiguous and broadly acceptable definition.[6] The census must make decisions, both in how it asks a question and in how it reports it. It decided, in its new series of reports on ethnicity, that (1) it would accept the individual's self-report as to his group, and (2) it would get at ethnicity by asking, "What is the respondent's origin or descent?"

To the student of ethnic affairs, these decisions are revealing and significant. Ethnicity, the census says, is not a matter of formal status so no one can be required to report it (though the administration of the laws does require that *other* people report it, as, for example, employers for employees, teachers for students). It is a matter of personal choice—no one will check on it and, presumably, no consequences will follow from giving an incorrect answer. Indeed, since it is a matter of the person's image of himself, who is to say what is an "incorrect" answer if someone answers Irish but is "really" mostly Slavic, or English but is "really" mostly German? Second the census avoids the term "ethnic" in the question and asks for origin or descent. Third, in view of the separation of church and state, the census will not only *not* ask for religion but will not make it possible for those who are members of an ethnic group which is also, confusingly enough, a religious group, to give that answer. "Jewish" or "Jew" is not on the card which the respondent is shown by the enumerator and from which he chooses an ethnic group. If the respondent nevertheless insists on reporting it, the census will not apparently record it—in any case, it does not report it.

Other approaches to all these questions are possible, even

in a country as close to us as Canada.[7] The Canadian census insists that each Canadian provide an answer on ethnic background, it will not accept the answer "Canadian," and it not only will accept Jewish as an answer for ethnicity, it will also accept it as an answer for religion. Whatever the failings of our own approach, for the first time we have some census figures on ethnicity and can report some key facts about white ethnic groups in the United States. The quality of this data is greatly superior to any we have had up to now, for the Current Population Survey sample includes approximately 47,000 households, while the public opinion survey which is the source of most of our recent information on ethnic groups covers generally a few thousand.

Surprisingly enough, out of a population of 204.8 million in March 1972, only 17.6 (8.6 percent) did not know their ethnic origin or did not report it. Another 24.8 million had, in the 1970 census, been counted as Negro, American Indian, Japanese, Chinese, and Filipino; 102.2 million listed themselves in one of eight major groups that were presented to them on a flash card. This left something like 60.2 million who must have reported they were of mixed or other origin. On this basis, we now know that 29.5 million Americans declare themselves to be of English, Scottish, or Welsh origin; 25.5 million German; 16.4 million Irish; 8.8 million Italian; 5.4 million French; 5.1 million Polish; 2.2 million Russian. There are 9.2 million who say "Spanish," which includes 5.2 million Mexican; 1.5 million Puerto Rican; .6 million Cuban; .6 million Central or South American; 1.2 million of "other Spanish origin."

One striking fact emerges immediately: Some of the ethnic groups that have been in the center of recent attention are quite small. The ethnic groups formed by Europeans of the "new immigration," that is, immigration after 1880 or 1890, include Italians, Poles, Russians, and other Eastern and Southern European groups. The Italians, by far the largest of them, form only 4.3 percent of the population—a lesser number than the "Spanish" of all origins. The Poles, the largest East European white ethnic group, consists of only 5.1 million, a mere 2.5 percent; the Russians only 2.2 million, a mere 1.1 percent. Certainly, against the 11 percent of the population that is black, these are tiny groups, and even if we put them all together, they number only 7.9 percent of the population. (It would hardly be proper to add all the Irish—about 8 percent of the population—since most are of the earlier immigration, and only part of them still form the neighborhood clusters that characterize the newer ethnic groups.) The Poles and Russians are even fewer than the figures given suggest, for they also include most of the Jews, whom the census, in its rigid effort to observe the constitutional prohibition of an "establishment" of religion, will not count.[8] (Jews make up—if we accept the estimates of Jewish organizations—about 2.9 percent of the population.)

We are a nation of immigrants, but we are not a nation of "ethnics," if we use this term as recent political discussion has done to include the descendants of recent European immigrant groups. Is, then, the recent attention and concern exaggerated? I believe not. For in some places—in particular in the large cities of the North and Midwest, where these groups are concentrated—recent ethnic groups form considerably larger proportions of the population, so that it is not incorrect to see some conflicts in these as primarily conflicts between blacks and perhaps Puerto Ricans on the one side, and recent white ethnic groups on the other. But there is a good deal of confusion in our usage of the term "white ethnic." In common usage it does not mean only the descendants of recent immigrant groups. It is also used as a surrogate for a number of other social categories of political significance. Thus, when we say "white ethnics," we also have in mind Catholics, who form between a quarter and a fifth of the American population, and a majority of the population in many Eastern and Midwestern cities. We have in mind broad occupational groups, the blue-collar and lower-middle-class workers. White ethnics thus seem to merge in common discourse with major religious groups such as the Catholic, and with major occupational groups, and indeed with "middle America" in general—an enormous, if vague, category. But is this usage legitimate? To answer this, we must have the answers to two further questions. First, are these other categories of religion and occupation distinctively associated statistically with the newer white ethnic groups, in such a way that the white ethnic of Eastern or Southern European background tends more than other Americans to be Catholic, urban, blue-collar and lower white-collar? And second, since the various roles—ethnic, religious, occupational—come in bundles, can we say that a political response is "ethnic," as against "religious" or "occupational"?

As to the first question: There is a close association between the new ethnic groups and two religions, Catholicism and Judaism. A study of the ethnic factor in Catholicism reports that about 20 percent of Catholics are of Italian background, 17 percent Irish, 16 percent German, 11 percent Polish, another 10 percent of other East European background. If we compare the figures on Catholic population from this sample survey with census figures given earlier, we will find that just about all the Italians and Poles but (surprisingly) only half the Irish, by the self-report of the census survey, are Catholic. The same study shows that these East and South European ethnic groups are concentrated in the Northeast and Middle West. One can safely assume a heavier-than-average concentration in the larger cities. Catholics very much reflect the occupational distribution of the country as a whole, except for a smaller number of farmers and a larger number of white-collar workers. Only the Poles and smaller East European groups, it seems, have a disproportionate concentration of blue-collar workers. Italians are at the national norm for blue-collar workers, Irish below the norm.[9] Jews are also concentrated in the Northeast and Midwest, and overwhelmingly in the larger cities. They have very few farmers, and few blue-collar workers.

The later ethnic groups are indeed not Protestant; they tend to be concentrated in the Northeast and Midwest; they probably are more distinctively to be found in the large cities; but they are not particularly—except for the East European Catholics—concentrated in blue-collar occupations.

Now to our second question: Which identity dominates in political action? When we consider the political orientations of any individual among these groups, it will not be easy to tell whether he responds as ethnic, as Catholic or Jew, Easterner or Midwesterner, as big-city-dweller, as blue-collar or white-collar worker. Undoubtedly, sophisticated analysis could throw some light on these matters. But I think our more general understanding of the processes of the way one's identity operates will be helpful. The fact is that people do have choices on how they identify themselves, they will choose different identities to emphasize in different settings, and the identities they choose over time will also change. Ethnic identities are alternatives to occupational, regional, religious, neighborhood, and other identities. Sometimes there is no conflict between these identities. The Polish blue-collar worker in an individually owned home in Detroit may see himself as Pole, Catholic, blue-collar worker, homeowner, Democrat, and defender of neighborhood turf, all in one. Occasionally, these different identities are in conflict with each other; when, for example, the Church or its representatives assert he must accept black children in the neighborhood school. It is this sort of problem that makes it so easy to label a political response "ethnic" and so hard to actually demonstrate it.

I take an agnostic view on the primacy of ethnic identity. I take an equally agnostic view on the primacy of class, or some other identity. It is an act of faith, not analysis, when one says that only class or occupational interests are real, and that when one acts on the basis of other interests or other identities, this is a matter of false consciousness, or a passing phase. In any case, since the different identities are bound together in statistical clusters and are carried by a single human being, it is very hard to separate out the various identities people carry and to relate the varying interests that activate an individual to specific identities.

The saliency of different identities changes over time, and perhaps the most significant reason that a "white ethnic" political reaction has become a focus of attention is that the saliency of ethnic identities has increased markedly since the middle Sixties: since, specifically, the Negroes became blacks, and the dominant tone of black political rhetoric shifted from emphasizing "we are like everyone else and want only integration," to "we are of course different from anyone else and want our proper share of power and wealth." This was a very striking change. It legitimated the raising of similar demands among such distinctive groups as Mexican Americans and Puerto Ricans. It helped create a Native American movement. And it helped, too, to legitimate the same kind of demands among white ethnic groups. These groups had emphasized: "We are like all other Americans, and even more so: We work harder, we are more patriotic, we are more anti-Communist." What the rise of a distinctive black political movement meant was that inevitably the question had to be raised, "Are we *indeed* like all other Americans only more so?" Or—another form of the question—"Is it to our *interest* anymore to emphasize this kind of identity rather than a separate identity as ethnics? If we are like all other Americans, then we bear the responsibility for slavery, exploitation, and imperialism. If we are, however, Poles, Italians, Jews, and the like, we have our own history of being exploited to refer to in protecting our position or extending it."

From the mid-sixties, I would say, the ethnic identity began to gain on the general American identity.[10] Indeed, the very term "American" became depreciated in the late 1960s. What happened, one wonders, to "I Am an American" days? We know what happened to committees on un-American activities. "Americanization" in schools is no longer considered desirable; indeed, Federal money is now available for just the opposite. In the ecology of identities, ethnic identities for certain purposes and in certain contexts began to serve individuals better than a general American identity.

Similarly, there was at least a stand-off between the desirability of asserting an occupational identity—blue-collar or lower-white-collar—as against an ethnic identity in cases of conflict. The prestige of the worker in American society began to drop after World War II and reached some sort of nadir in the 1960s, as evidenced, for example, by the fact that working-class styles of life and working-class ambitions were no longer portrayed with sympathy in the mass media. The labor unions lost their attractiveness to intellectuals and there was little to be found in the mass media or popular press that presented them sympathetically. To the educated, the trade unions were tarred with the charge of discrimination against blacks. When one spoke as an ethnic, then, one was in a sense freed from carrying the burden of defense of trade-union discriminatory behavior, and one emphasized instead that one was defending something positive—the neighborhood, the neighborhood school, some distinctive neighborhood, and family values. One was not, as an abstract American white, simply *excluding* the black.

Thus the question of whether there is such a thing as a white ethnic political reaction is, as we see, a complex one—even before one gets to any political data. An answer must involve some consideration of how these identities are formed in America, the circumstances under which they are rejected or considered somewhat disreputable, the circumstances under which they are rehabilitated and given some aura of respectability. I believe the circumstances of the past eight or nine years have, for many persons of the later ethnic groups, tended to re-habilitate ethnic identities.

What are the specific interests, however, that have encour-aged this rehabilitation, and to what extent are these "ethnic"?

There is the common view that the later ethnic groups share some particular deprivation, are concentrated in inferior occu-pations, make less money, have less access to higher education than "old Americans," and less than the descendants of the old immigrant groups. The Irish, while an old immigrant group parallel in time of migration to the German, are often included among the American ethnic groups that are deprived in rela-tion to old American or older, white Protestant ethnic groups. It is, perhaps, the political skill of the Irish which makes it possible for an old immigrant group to maintain its relation-ship to the new immigrant groups, as "one of them." More likely, it is the common Catholicism which makes this possible. But are these groups really deprived in relationship to white, Anglo-Saxon Protestant Americans? Largely, no, as far as one can tell. We are not surprised to learn that the Poles and Russians in the census survey are disproportionately concen-trated in professional, technical, and kindred occupations, and among managers and administrators (28.1 percent in these two categories for Poles, 38.3 percent for Russians), figures above that for English, Scottish, and Welsh (26.8 percent), and close to that for Germans (31.3 percent); for the Poles and Russians, we know, include the upwardly mobile Jews. But more Italians, too, are to be found in the two top occupa-tional categories (31.0 percent) than English, Scottish, and Welsh. If the Italians, the only large new group that is not mixed with others in the census survey, are representative of the new ethnic groups, they have not done badly. If we look at median income figures, we will find—again leaving out the Russians and Poles, who come at the top, owing to the Jewish admixture—that the Italian median income ($11,646) is higher than that of English, Scottish, and Welsh ($11,345), and of the German ($10,977). As for education, all the groups are bunched together when one excludes Russians.[11] The deprivation, then, is neither economic nor occupational nor educational when one compares the later ethnic groups with old Americans and older ethnic groups. Of course, one can always find a group in relation to whom one *is* deprived. One study of Detroit which breaks down the white Protestant group into specific denominations discovers that even the Jews are surpassed by Congregationalists, both in income and education.[12]

Clearly, if part of the explanation for a white ethnic political movement is economic, it is not because any particular economic deprivation can be pointed to as affecting these groups.

These are, of course, crude and initial comparisons. Have the later ethnic groups suffered, in contrast to their position ten years ago, more than other groups? Perhaps. But an analysis by Andrew Greeley argues quite the opposite. He writes that:

. . . The evidence in NORC and SRC composite samples leaves little doubt that the Irish Catholics, Germans, Scandinavians, and Italians have moved rapidly into the upper-middle class of American society during the last two decades. Furthermore, despite fears that these groups would become alienated, there is no evidence that between 1950 and the late 1960's any of them suffered appreciable decline in their relative incomes. Whatever the inflationary income squeeze of the 1960's may have been, it does not seem to have affected these ethnic groups disproportionately. . . . The Polish-Catholic situation is uncertain. While there is some evidence that the Poles are leaving their blue-collar status behind and are moving upward relatively as well as absolutely, the speed of their movement seems to be considerably less than that of the Italians. . . . The mobility system did not work at all for the Blacks between the 1950's and the 1960's. [As we have seen, it seemed to begin working in the later 1960s—see Chapter 2, pp. 41–43.] It seems to work to the disadvantage of the Irish Protestants and Other Protestants or "Americans," who have slipped . . . behind. . . . If there are any ethnic groups, then, that have suffered in the last two decades, they are the older groups, those that have been here since the beginning of the republic.[13]

The second set of deprivations of the later ethnic groups that might explain a political reaction and that we should consider are cultural deprivations—the requirement to give up distinctive cultures, languages, aspects of religion, life-styles, all in the service of Anglo-conformity. Perhaps what is really bothering the later ethnic groups and leading them to demand political redress is a cultural loss rather than an economic or occupational one. They see black studies programs everywhere, Chicano and Puerto Rican studies programs widespread, Asian studies in those colleges where there are substantial numbers of these groups. But where are Italian or Polish or Ukrainian studies programs? Where is the recognition of their cultural heritage?

The sense of cultural loss is a complex thing. It may consist literally of a concrete loss: the desire to speak and maintain a language and hand it on to one's children, and the pain felt at the inability to use it to serve the ordinary business of work and life. A rather different feeling may be involved, in which one does not oneself wish to maintain language and custom but believes others in the group who wish to should have that right and some public support in doing so. It may consist only of the insistence that the worthiness of the cultural tradition be recognized even though hardly anyone in the group wishes to practice it.

One dimension in an ethnic concern for one's culture is based on one's personal involvement in it, ranging from intense to distant or nonexistent. A second dimension is political: the feeling that if any culture is to be recognized, this one should be, also. One aspect of this is concern over the degree of respect given the group's culture. Thus the resistance to the Polish joke, to the mass media identification of Italians with crime, to characterizations of Jews that suggest anti-Semitism may have nothing to do with Polish, Italian, or Jewish culture. It has to do only with the defense of the respectability of the group.

Thus the question of an ethnic cultural loss, and a cultural basis for political action, is complex. There may be attachment to a concrete language and cultural style which is being inhibited by patterns of public education, work requirements, and so on. There may be the sense of political disadvantage. There may be the deprivation that comes from public disrespect or depreciation. There are undoubtedly other dimensions. Within each group, different matters agitate and concern different people. And different groups as a whole have different concerns. Spanish-speaking groups, with language difficulties, may insist on some means by which public agencies respond to their special language problem, and one way is by publicly accepting the use of Spanish. Other groups simply want their fair share of public funds for ethnic activities. Other groups are worried more about respect. Ethnic leaders may seize on any of these concerns, even contradictory ones, in order to legitimate their role and to demonstrate that ethnic organization is necessary.

The specific demands that members of a single group make also differ in objective. Should one recognize black English so that blacks who speak it should learn standard English *better?* Or should one recognize it to legitimate its use in various parts of the culture (beginning with the schools) as an acceptable alternative? Similar questions arise with bilingual programs. The most widely accepted rationale is probably that they permit those who speak a non-English language to learn English better. But a second rationale is that non-English speakers should not be penalized in practical matters such as school grades because they do not know English better. A third objective is that they should not lose respect for their own cultural tradition. A fourth is that a non-English speaker should be able to participate in the culture in the same way that speakers of English do, in its educational institutions, its work opportunities, its political system. A fifth is that the culture and language of each group must be maintained in all its fullness. These are all quite different objectives and very different bilingual programs might well be required to attain each of them.

Having developed this array of possible "ethnic" demands—and we have scarcely exhausted the possibilities—which among them concern the later white ethnic groups? Do they feel, as Michael Novak and many others do, the loss of a distinctive culture, that they were forced to give up in order to achieve position in American society? [14] One suspects this is a point of view which is more likely to be found among intellectuals than other people. It is hard to mourn the loss of a lost culture when one does not possess or recall it. It is not my impression that Polish and Italian Americans in general mourn the loss of the whole, complete, integrated—and poverty-stricken—lives of Polish and South Italian villages, even though intellectuals who return to them and study the sources and experience what part of this life still continues may well feel a sense of loss. Some Jewish intellectuals experience a sense of loss when considering their past. Isaac Bashevis Singer, the recorder of the *shtetl,* and Bernard Malamud, who writes about the *shtetl* transplanted, are very popular, and young American-born Jewish writers have begun recently to recreate the Jewish East European past in novels.[15] The Jewish students who in the 1960s began to search for a Judaism or Jewishness that would be relevant to them also echoed Michael Novak's distress. Thus they criticized their parents—the "Uncle Jakes," the Jewish parallel to the Uncle Toms—for having given up Jewish culture for success, for not doing enough to push the American government to save the Jews of Eastern Europe during World War II. They began to experiment with Jewish life-styles. We have seen a similar phenomenon among Japanese and Chinese Americans, who also criticize their parents for not doing enough to fight the relocation camps of World War II or the anti-Chinese discrimination of the past. And there are similar developments among students in a number of East and South European ethnic groups.

And yet, when one surveys the examples of political activity among these groups, it is hard to see that the mourning or search for a lost culture is a powerful stimulus affecting large numbers. One sees in many individuals a criticism of the Americanization of the first and second generation, a kind of "third generation return," to use the formulation of Marcus Hansen's provocative essay asserting that such a phenomenon could be seen among the earlier ethnic groups. But it is my judgment that the issue raised is, in large measure, a symbolic one. The demand for recognition of a distinctive culture does not indicate the desire to recreate it and scarcely the desire to maintain it. Rather, individuals in each group make a claim to a fair measure of respect as full members of the emerging American social order. Symbolic demands are by no means unimportant, but it is a fact they can be met by rather simpler adjustments than demands that arise from a concretely different culture. The problems raised by the Spanish-speaking are serious ones. The problem about what to do for demands for Swahili in the schools, or for programs in certain places for Polish or Italian or Jewish studies, is a rather different matter. When I say these demands are in large measure symbolic, I suggest that a symbolic response is sufficient to meet them. To really teach all blacks Swahili or all students of Italian or Jewish or Polish background the language of their forefathers and something of their culture would be a quixotic enterprise, but it is not a problem to establish the *possibility* for those inclined to learn something and for some to become experts. I conclude that, insofar as the political demands of the recent ethnic groups relate to the specific culture of specific ethnic groups, culture understood as language and custom and history, they are basically symbolic demands. The recent ethnic groups—this is one observer's judgment, but the fact is that there is as yet little evidence on one side or the other—do not want the maintenance or revival of traditional culture in any full or concrete way. As part of their search for equality and respect, they want to see these cultures recognized. Here, indeed, "tokenism" is all that is being demanded and all that is necessary.

There is a third possible basis for a distinctive political concern among later ethnic groups: When there is a peril to the homeland, specific ethnic interests and attachments may become truly dominant among a group. The main current examples are the impact of the threat to Israel—a special kind of homeland—on American Jews, of the Near East crisis on immigrants from Arab countries, and of the conflict over Cyprus on American Greeks. In the 1940s and 1950s, Communist domination of East European countries had the same effect on groups from those countries here. But a compromise seems to have been worked out between Communism and nationalism in Eastern Europe, and that, as well as the passage of time, has quieted concerns among most of these groups for their homelands. The Baltic peoples remain unreconciled. This is a recurrent basis of ethnic political action, even though the only powerful examples today are Jews, Arabs, and Greeks.

If there are no distinctive economic deprivations among the later ethnic groups, if there are no strong concerns over cultural deprivations and the maintenance of the specifically cultural aspect of the ethnic heritage, if there is—except for Jews —less concern over threatened homelands, is there any basis at all for a distinctive political movement among white ethnics? In recent years, that distinctive basis, it has been asserted, is to be found in the peculiar racism of these groups, a racism fed by the reality of conflict over jobs at the lower level in the past, over jobs at the middle and higher levels today, and that becomes particularly emphatic when there is a threat of black entry into a white working-class or middle-class neighborhood.

Thus the explosive issues of jobs, schools, and residential segregation arouse the hostility of white ethnics against blacks, and create a distinctive political interest to which racist politicians such as Wallace can appeal and which liberal Democratic politicians must confront.

We have a good number of surveys on the extent and depth of prejudice among the later ethnic groups, and it appears, surprisingly, that they are less prejudiced against blacks, and more supportive of their advance, than white Protestants, who, in general, represent descendants of old settlers and older ethnic groups. Two pieces of evidence: A Gallup survey of the late Sixties asked: Is integration moving too fast? The national average saying yes was 48 percent. Among blue-collar Catholics, this dropped to 42 percent. Among white, blue-collar Protestants, this rose to 57 percent. Thus blue-collar Catholics were *less* prejudiced by this measure than the national average. NORC has developed an integration scale based on 7 items. A 1970 test shows that every ethnic group is more pro-integration than white Anglo-Saxon Protestants. By degree of support for integration, the order runs: Jews, Irish, German Catholic, German Protestant, Scandinavian, Italian, Slavic, and the WASP.[16] Another analysis of national surveys reports, ". . . in large metropolitan areas, Protestants are more likely to say they definitely would move or might move if a Negro family moved next door than would Catholics." [17]

Of course, survey results are in no sense decisive, but it is hard to believe that white ethnics would try *harder* to conceal racist sentiments than white Protestants.

But there is indeed a special edge, a special tension, in the relationships between white ethnics and blacks and rising minorities. Undoubtedly this arises in large part, out of general racist sentiments. In more substantial part it arises out of the conflict over jobs. The later white ethnic groups are strongly represented in the trade unions and the areas of construction, transportation, and manufacturing in which trade unions are powerful. Their progress has been based in large measure on union strength. They are committed to the rules and principles that union strength has established, for example, the overriding role of seniority in determining opportunities for advancements and protection from layoffs and dismissals. They see these principles now challenged and restricted by affirmative action and quotas for blacks and other groups.

It is understandable that there should be this kind of pragmatic, interest-based conflict. But how does it explain the fact that the resistance of the later ethnic groups to entry of blacks into their residential areas, of black children into the schools of their areas, and, most markedly, to the transportation of their children to attend schools in black areas is so fierce? Is there something special and distinctive about the later white ethnic groups that explains this resistance? One would expect that, if pragmatic concerns were dominant, the resistance to the entry of blacks into jobs that were held for the members of these groups and their children, and to the more rapid advancement of blacks than of members of this group, would arouse the greatest resistance. But the movement of blacks into jobs, under affirmative action requirements, court orders, consent agreements, and the like, has up to the present met much less resistance than either residential movement or racial balancing in the schools.

One reason of course the later white ethnic groups seem to resist such changes more strongly is simply that they are in the path of black residential movement (as against earlier ethnic groups who are already further out in the suburbs) and their children are subject to the requirements for racial balancing, since they are still more heavily concentrated in the central cities. But there is something else which gives a specific edge to the relations between blacks and some other minority groups and the later white ethnic groups involved. There are, first, significant differences of life-style, of "culture," not in the sense that there are distinctive white ethnic cultures among the later groups that they particularly wish to defend, but in the sense that there are ways of life that come into conflict. There is, second, the personal impact of the comparison of experience: the members of later white ethnic groups feel that they are unfairly being asked to take on a burden. This raises the issue of unfairness, which, once raised in a world where ideas of equity and equality become increasingly dominant, has enormous power to move people. Neither of these themes is easy to concretely grasp or present. A good part of the salience that each possesses is based not on differences of fact but on differences in perception. Thus, it is undoubtedly true that more black families today are broken; there is a higher rate of illegitimacy, and there is a higher rate of crime among the young, than in the later white ethnic groups. Whether there would be such differences if the later white ethnic groups suffered the same job and residential situation as poor blacks in inner-city areas is another question. Indeed, whether there were such differences when the later ethnic groups first settled in the cities, we do not know. We do know there was a great deal of prostitution, crime, runaway husbands, broken families, incest. There is disturbing evidence that the black family was stabler in earlier periods in the Northern cities than it is today, even though, by the measures I presented earlier, the economic situation has improved. Perhaps we may ascribe this decline in family stability to the mixed blessings of welfare, which the white ethnic groups, in a pre-welfare society, were spared.

But whatever the causes to which we ascribe the family breakup and social disorder among blacks in poor areas in Northern cities today, the differences between black areas today and the neighboring areas inhabited by the later white ethnic groups are great. The latter are scenes of a marked social order: stable neighborhoods, with children succeeding parents in the same area, strong organizations centered around the church, formal ethnic associations or patterns of informal ethnic association, the local political organization, the trade union, the local small businesses of members of the group, which serve as much for socialization as for ordinary business. The black areas are strikingly different: high proportions of female-headed families, an uncertain place for adult men, crime among the young frightening the inhabitants and warning away any potential passers-by. These are realities, and it is hardly likely they would not lead to resistance in the white ethnic neighborhoods to the entry of blacks or resistance to having their children sent into the black areas. Admittedly, the perceptions are crude and stereotyped, and the stable, black working-class and middle-class family seeking better housing and education are outraged—and legitimately—at being taken as potential menaces rather than for what they are.

But the differences between the groups are real, and they extend to dress, talk, gesture, walk, public behavior. The simple characteristics of a different subculture may be taken by white ethnics as a symbol of what they fear—threats to safety, morality, neighborhood stability, and legitimate order and authority. Two sociologists studying mixed working-class neighborhoods in Chicago give, to my mind, a convincing description of how these differences work their way into every aspect of social intercourse and lead to hostility. Gerald Suttles summarizes the characteristics of two ethnic groups living on the West Side of Chicago as follows:

Among the Italians, people from all walks of life are drawn together in a well-knit series of peer groups that range from childhood to the upper realms of adulthood. Both the "church people" and the racketeers are bound together in a common collusion of "impression management" [the term is Erving Goffman's] and are equally safe in each other's presence. Local business establishments, street corners, and other public facilities are categorized according to their proper "hangers-on," the license they may enjoy, and the behavior appropriate to "outsiders."

At the other extreme are the Negroes, who, like the three ethnic groups [there are also Mexican and Puerto Rican areas], form a small but compact residential group. They . . . remain the most estranged from one another. Anonymity and distrust are pervasive, and well-established peer groups are present only among the adolescents. Sometimes those residents who are most "respectable" carry on a futile and divisive attack on those who are not so respectable. . . . Local businesses, street locations, and other spatial settings, with the exception of adolescent hangouts, are not well-differentiated according to who can be there and what behavior is required of them.

Here is Suttles' perceptive discussion of small businesses in such areas:

Many business establishments are so thoroughly acknowledged the property of a single minority group that customers outside the group seem like intruders. When I first went into the Addams area, I entered several places where I was asked "Whatta you want?" as if I were lost. What I did not know was that these places are thoroughly tailored to the needs and personal peculiarities of a small network of friends within a single minority group. To them my presence was totally inexplicable. At worst, I could be a policeman, some sort of city inspector, or a trouble-maker. Whatever had led me there could not have been their products because they were too expensive, of questionable quality, badly displayed, and usable only to an ethnic group quite other than my own.

All the ethnic groups in the area have such establishments. . . .

But in the black areas, these establishments are few and are run by members of other ethnic groups—in part because of the well-known absence of blacks in small business,[18] in part because in this area they live in a housing project, and housing projects do not have facilities for small businesses. But it is common for white ethnic groups to live in privately owned dwellings, blacks in housing projects—half the people living in housing projects are black.

Thus blacks are treated as intruders in businesses catering to an ethnic group. Even when a business opens up to outsiders—as had an Italian restaurant—blacks are considered harmful because to the proprietor they threaten its ethnic authenticity and respectability. Blacks sense the hostility, and they return it. The white-run businesses in their own area they regard with a similar hostility:

. . . It is hard for the Negroes to explain the presence of the white operators except as rank opportunism which takes advantage of their own weakness. The store suffers greatly from shoplifting and upaid bills, and has been broken into. . . . Probably only a small proportion of Negroes engage in these forms of predation. Nonetheless, those who do are not much disapproved of, and their actions are regarded as only "fair." [19]

Quite similar in tone is William Kornblum's description of another predominantly white ethnic neighborhood, with blacks and Mexicans, in South Chicago. Here the white ethnics are of various Slavic groups. Here, too, the blacks are seen as a threat. The young blacks come into the steel mills and work under white ethnics who have worked their way up over the years to positions of foremen, and approach their work with a very different attitude from that of their Slavic predecessors. "Milan," a Serbian foreman, Kornblum writes,

has been disturbed by two of the black laborers. . . . Both are Chicago-born, and accustomed to street life, which they complain they miss when they work evenings or nights. . . . Both are frequently absent on week-end turns [shifts] and on other occasions have come to work noticeably drunk. . . .

[Milan] calls them Mau Maus and says he intends to get them out of No. 3 Mill even if the higher management will not back him up.

Kornblum describes an incident in which one black worker has come late and reports his friend will come later. The foreman tells him his friend will be sent home. The worker protests, and the foreman threatens to send him home, too, at which point he says angrily, "No white motherfucker is going to talk to me like that." And so on, leading to a discipline report by the foreman, and a grievance filed by the black worker. Milan says, "One of these young shines is going to kill a foreman someday soon. You can see it in their eyes, those black bastards. It used to be nice to work here, now with all these young niggers coming into the mill you never know what's going to happen. You take your life in your hands when you tell them to do something." [20]

The differences between the Italian and East European neighborhoods and the black neighborhoods expanding into them cannot be wished away: the first emphasize strongly a neighborhood-centered, family-centered, job-centered life, respect for authority, close attachment to local institutions. Most intrusions would in any case be seen as a threat. When this style of life meets with the radically different styles of life of poorer black communities, one has suspicion and hostility. It may well be true that black styles of life respond to black experience, that is, oppression and discrimination, but that will not ease matters when white ethnics one generation removed from their immigrant fathers and blacks one generation removed from the Southern caste system come into contact.[21]

The significance of family and neighborhood in white ethnic thinking is suggested in a recent manifesto of a group of white ethnic leaders launching an "Ethnic Millions for Political Action Committee." For example:

EMPAC! represents a people-centered approach to politics. Of any political or economic program, it asks: (1) Is it good for families and neighborhoods? (2) Are the needs of all affected groups represented in an articulate and effective way among those who make decisions?

EMPAC! supports political and economic initiatives organized around the following points:

(1) A national policy for strengthening family life in America, especially in regard to children and the aging, and to new roles for women. From the beginning, a high proportion of white ethnic women have held occupations outside the home; yet the values of homemaking and family life are also deeply cherished. Tax incentives for cooperative family child care are needed by the extended family and its modern substitutes.

(2) A national policy for strengthening neighborhood life and

services in America. Neighborhoods need long-range stability, openness, cultural satisfaction, peace, and mutual aid. Family investment in a home and business should be guaranteed by the government against social change, as bank deposits are now guaranteed.

(3) A national policy for strengthening social identity in America and a national effort to construct an accurate history of American diversity—in schools, texts, university research, the media, and other institutions. Rootlessness and atomization result from the neglect of reality.[22]

The organizers of this group are liberal and want an alliance with blacks, Spanish-origin groups, and labor. But to some readers the emphasis on "family" and "neighborhood" will sound like "code words" inserted to summon up anti-black sentiments. Some words are indeed code words, but we have to be careful not to ignore the manifest content of words in favor of their presumed latent content. At the base of the conflict, I believe, are less racist attitudes as such than perceived differences in behavior, differences which are seen as threatening. Gary Jacobson has analyzed the reports on the 1972 election, trying to estimate the weight of racism in the flight of Democratic voters from McGovern. He writes:

. . . relevant to the question of how racial attitudes may have influenced defectors from the Democratic ranks in the 1972 Presidential election are the studies of Milton Rokeach and his followers. Rokeach's theory, generally confirmed in both laboratory and real-world situations, is that, unless racial segregation is socially enforced, it is not really *race* that most people discriminate. Rather they discriminate according to beliefs. That is, feelings of attraction or repulsion between people (and behavior in line with these feelings) depend more on congruency or dissimilarity of belief than upon race. People in experimental situations will choose to associate with persons of a different race who appear to share their beliefs much more frequently than they will choose to associate with persons of the same race who have opposing belief systems. . . .

193

The implications of this are enormous. It could be that at present the real source of prejudice is not race at all, at least for most people; perhaps they are *not* deceiving themselves or lying to conform to expected social norms when they express unprejudiced racial attitudes. They really reject Blacks and other racial minorities not because of their skin color, or because they believe that these minorities are by nature inferior, but because they believe them to act according to belief systems contrary to their own. Talk about "law and order," "welfare ethic," "immorality," or "crime in the streets" may not really be code words for racism at all, but rather they are the cause and essence of the racial attitudes themselves. They may be the real issues after all, even though in a circular way they may lead to behavior that on the surface can be interpreted as racist in the usual sense of the word.[23]

There is a second factor that introduces a special tension between the later white ethnic groups and blacks: the comparison of experience. From the point of view of the white ethnics, they entered a society in which they were scorned; they nevertheless worked hard, they received little or no support from government or public agencies, their children received no special attention in school or special opportunity to attend college, they received no special consideration from courts and legal defenders. They contrast their situation with that of blacks and other minority groups today and see substantial differences in treatment. They consider themselves patriotic and appreciative of the United States even though they received no special benefit. They look at the minority groups and find them abusive of the state though they do receive special benefit. This may be a crude and unfair comparison; after all, the blacks were brought in chains as slaves and the whites came as free men, blacks have continually dealt with the most severe and unbending prejudice, whereas that met by immigrants was mild and scarcely to be found after the second generation, and we could continue the comparison. But the perception cannot be dismissed as false either, and however we disagree over how "true" it is historically, it plays a great weight in politics and in the belief of white ethnic groups that they are subject to unfair policies.

Nothing is so powerful in the modern world as the perception of unfairness. Perhaps it is inevitable that wherever there are distinguishable groups, one group will feel it is unfairly treated in contrast with another. A strong sense of unfairness is widely found among the later white ethnic groups when they consider the policies I have described in Chapters 2, 3, and 4; this, along with the perceived differences in culture, contributes to the special edge in the relationship between these two elements of the American population.

CHAPTER

6

Morality, Politics, and the Future of Affirmative Action

A NEW COURSE in dealing with the issues of equality that arise in the American multiethnic society has been set since the early 1970s. It is demonstrated that there is discrimination or at least some condition of inequality through the comparison of statistical patterns, and then racial and ethnic quotas and statistical requirements for employment and school assignment are imposed. This course is not demanded by legislation—indeed, it is specifically forbidden by national legislation—or by any reasonable interpretation of the Constitution. Nor is it justified, I have argued, by any presumed failure of the policies of nondiscrimination and of affirmative action that prevailed until the early 1970s. Until then, affirmative action meant to seek out and prepare members of minority groups for better jobs and educational opportunities. (It still means only advertising and seeking out in the field of housing.) But in the early 1970s affirmative action came to mean much more than advertising opportunities actively, seeking out those who might not know of them, and preparing those who might not yet be qualified. It came to mean the setting of statistical requirements based on race, color, and national origin for employers and educational institutions. This new course threatens the abandonment of our concern for individual claims to consideration on the basis of justice and equity, now to be replaced with a concern for rights for publicly determined and delimited racial and ethnic groups.

The supporters of the new policy generally argue that it is a temporary one. They argue (or some do) that consideration of race, color, and national origin in determining employment and education is repugnant, but it is required for a brief time to overcome a heritage of discrimination. I have argued that the heritage of discrimination, as we could see from the occupational developments of the later 1960s, could be overcome by simply attacking discrimination. The statistical-pattern approach was instituted *after,* not before, the remarkably rapid improvement in the black economic and occupational position in the 1960s. I have argued that the claim that school assignment on the basis of race and ethnicity is only temporary is false, because the supporters of such an approach now demand it whatever the circumstances, and the Constitution is now so interpreted that it can be required permanently.

We have created two racial and ethnic classes in this country to replace the disgraceful pattern of the past in which some groups were subjected to an official and open discrimination. The two new classes are those groups that are entitled to statistical parity in certain key areas on the basis of race, color, and national origin, and those groups that are not. The consequences of such a development can be foreseen: They are already, in some measure, upon us. Those groups that are not considered eligible for special benefits become resentful. If one could draw a neat line between those who have suffered from discrimination and those who have not, the matter would be simpler. Most immigrant groups have had periods in which they were discriminated against. For the Irish and the Jews, for example, these periods lasted a long time. Nor is it the case that all the groups that are now recorded as deserving official protection have suffered discrimination, or in the same way.

The Spanish-surnamed category is particularly confused. It is not at all clear which groups it covers, although presumably it was designed to cover the Mexican Americans and Puerto Ricans. But in San Francisco, Nicaraguans from Central America, who were neither conquered by the United States nor subjected to special legislation and who very likely have suffered only from the problems that all immigrants do, given their occupational and educational background, their economic situation, and their linguistic facility, were willy-nilly swept up into one of the categories that had to be distributed evenly through the school system. The Cuban immigrants have done well and already have received special government aid owing to their status as refugees: Are they now to receive, too, the special benefit of being considered Spanish-surnamed, a group entitled to the goals required in affirmative action programs?

The protected groups include variously the descendants of free immigrants, conquered peoples, and slaves, and a single group may include the descendants of all three categories (e.g., the Puerto Ricans). Do free immigrants who have come to this country voluntarily deserve the same protected treatment as the descendants of conquered people and slaves? The point is that racial and ethnic groups make poor categories for the design of public policy. They include a range of individuals who have different legal bases for claims for redress and remedy of grievances. If the categories are designed to correct the injustices of the past, they do not work.

They do not work to correct the injustices of the present either, for some groups defined by race and ethnicity do not seem to need redress on the basis of their economic, occupational, and educational position. The Asian Americans have indeed been subjected to discrimination, legal and unofficial alike. But Chinese and Japanese Americans rank high in economic status, occupational status, educational status. (This does not prevent members of these groups from claiming the benefits that now accrue to them because they form a specially protected category under affirmative action programs.) If they were included in the protected category because they have faced discrimination, then groups in the unprotected categories also deserve inclusion. If they were included because they suffer from a poor economic, occupational, and educational position, they were included in error. So if these ethnic and racial categories have been designed to group individuals with some especially deprived current condition, they do not work well. Just as the Chinese and Japanese and Indians (from India) do not need the protection of the "Asian American" category, the Cubans do not need the protection of the Spanish-surnamed category, and middle-class blacks the protection of the Negro category, in order to get equal treatment today in education and employment. The inequalities created by the use of these categories became sharply evident in 1975 when many private colleges and universities tried to cut back on special aid for racially defined groups, who did indeed include many in need, but also included many in no greater need than "white" or "other" students. But the creation of a special benefit, whether needed or not, is not to be given up easily: Black students occupied school buildings and demanded that the privileges given on the basis of race be retained. This is part of the evil of the creation of especially benefited ethnic and racial categories.

The racial and ethnic categories neither properly group individuals who deserve redress on the basis of past discriminatory treatment, nor properly group individuals who deserve redress on the basis of a present deprived condition. The creation of such specially benefited categories also has inevitable and unfortunate political consequences. Groups not included wonder whether it would not be to their benefit to be included. In India, whose history and circumstances are entirely different from ours, the scheduled castes, and then tribals, were given special rights in employment and education. Very likely one can make an excellent case for these "reservations," as the Indian quotas are called. But other groups have tried to qualify for these rights, to the point where an Indian state court had to rule that no more than 50 percent of the population could be so included—presumably, at that point a new minority was created that was being discriminated against. There are already cases of individuals who redefine themselves in this country for some benefit—the part American Indian who becomes an Indian for some public purpose, the person with a Spanish-surnamed mother who now finds it advantageous to change his or her name; conceivably the black who has passed as white and who may reclaim black status for an educational or employment benefit. We have not yet reached the degraded condition of the Nuremberg laws, but undoubtedly we will have to create a new law of personal ethnic and racial status to define just who is eligible for these benefits, to replace the laws we have banned to determine who should be subject to discrimination.

The gravest political consequence is undoubtedly the increasing resentment and hostility between groups that is fueled by special benefits for some. The statistical basis for redress makes one great error: All "whites" are consigned to the same category, deserving of no special consideration. That is not the way "whites" see themselves, or indeed are, in social reality. Some may be "whites," pure and simple. But almost all have some specific ethnic or religious identification, which, to the individual involved, may mean a distinctive history of past—

and perhaps some present—discrimination. We have analyzed the position and attitudes of the ethnic groups formed from the post-1880 immigrants from Europe. These groups were not particularly involved in the enslavement of the Negro or the creation of the Jim Crow pattern in the South, the conquest of part of Mexico, or the near-extermination of the American Indian. Indeed, they settled in parts of the country where there were few blacks and almost no Mexican Americans and American Indians. They came to a country which provided them with less benefits than it now provides the protected groups. There is little reason for them to feel they should bear the burden of redressing a past in which they had no or little part, or assisting those who presently receive more assistance than they did. We are indeed a nation of minorities; to enshrine some minorities as deserving of special benefits means not to defend minority rights against a discriminating majority but to favor some of these minorities over others.

Compensation for the past is a dangerous principle. It can be extended indefinitely and make for endless trouble. Who is to determine what is proper compensation for the American Indian, the black, the Mexican American, the Chinese or Japanese American? When it is established that the full status of equality is extended to every individual, regardless of race, color, or national origin, and that special opportunity is also available to any individual on the basis of individual need, again regardless of race, color, or national origin, one has done all that justice and equity call for and that is consistent with a harmonious multigroup society.

Each of the policies we have discussed of course raises special problems. Inclusion in employment goals and quotas is clearly a positive benefit for individuals in the benefited groups, an actual loss for the others. As a result, these benefits will be defended most fiercely. It is less clear what the benefit is in school desegregation. It is obviously considered no benefit at all but an actual loss by some of the populations involved, white and Asian American, may well be seen as a loss by many of the Spanish-surnamed groups involved, and is even seen as a loss by many of the blacks. Residential distribution is an even more ambiguous case. It is a clear gain if it means access to better housing and communities with better services. But it may imply only the dubious benefit of housing in a project or low-income section of the suburbs, rather than a project or a low-income section in the central city; and if we consider this concrete reality, many for whom such a policy is designed may well see it as a disadvantage, too.

These policies are based on two equally inadequate views of the nature of racial and ethnic groups in the United States. First, they assume that these groups are so easily bounded and defined, and so uniform in the condition of those included in them, that a policy designed for the group can be applied equitably and may be assumed to provide benefits for those eligible. The fact is that, for many of the groups involved, the boundaries of membership are uncertain, and the conditions of those included in the groups are diverse. Furthermore, it clearly does not serve the creation of an integrated nation for government to intervene in creating sharper and more meaningful ethnic boundaries, to subdivide the population more precisely than people in general recognize and act on. The fact is, a good deal of integration—taken in the fullest sense of the term—has gone on in this country. The process will not be aided by trying to fix categories for division and identification and then make them significant for people's fates by law. We are not a nation such as Belgium, which can draw a geographical line across the country and pronounce one side Flemish and the other French (and even Belgium has a serious problem in considering what to do about mixed—that is, "integrated"—Brussels). Nor do we want to become such a nation, in which our people live within ironclad ethnic and racial divisions defined by law.

These policies make a second and opposite error, and that is to ignore the reality that some degree of community and fellow-feeling courses through these groups and makes them more than mere assemblages of individuals. We have seen how school desegregation policies have taken a positively hostile attitude toward any expression of such a group reality, how residential distribution policies assume that any community is a "ghetto," imposed from without rather than chosen from within. Even statistically oriented employment policies also ignore certain realities of community. Racial and ethnic communities have expressed themselves in occupations and work groups. Distinctive histories have channeled ethnic and racial groups into one kind of work or another, and this is the origin of many of the "unrepresentative" work distributions we see. These distributions have been maintained by an occupational tradition linked to an ethnic community, which makes it easier for the Irish to become policemen, the Italians fruit dealers, Jews businessmen, and so on. None of us would want these varied occupational patterns maintained by discrimination. Nor, however, should we want to see the strengths provided by an ethnic-occupational link—strengths for the group itself, and for the work it contributes to society—dissipated by policies which assumed all such concentrations were signs of discrimination and had to be broken up. A rigorous adherence to requirements of no discrimination on grounds of race, color, and national origin would weaken these concentrations and offer opportunities to many of other groups. A policy of statistical representation in each area of employment would eliminate them—but that would be to go beyond the demands of justice and equity.

Thus policies of statistical representation in employment, education, and residence insist that it is possible to divide the racial and ethnic groups with precision and assign them on the basis of past discrimination and present circumstance to a class for which a strict statistical parity must be required, and a class which does not warrant this protection (if protection it is—as we saw, in some cases, even the protected groups look dubiously on what is proposed for their putative benefit). But on the other hand, these policies insist that despite the precision with which these groups may be defined and discriminated, none of them may exist in any group or corporate form even if this is a matter of their own choosing. In contrast, the emergent American ethnic consensus we described in the first chapter of this book insisted that the group characteristics of an individual were of no concern to government, that it must take no account of an individual's race, color, or national origin. And on the other side the consensus insisted that any individual could participate in the maintenance of a distinctive ethnic group voluntarily, and that government could not intervene to break up and destroy these voluntary communal formations. Finally, I argued in the first chapter, the American ethnic consensus would not accept these voluntary racial and ethnic formations as component parts of the American polity. There were to be no group political rights in addition to the rights of the individual and of the constituent states. This element, too, of the consensus is in process of being subverted by the emergence of a required statistical presentation of some racial and ethnic groups in key areas of life, because to ascribe rights on the basis of ethnic group membership inevitably *strengthens* such groups and gives them a greater political role.

How have policies which so sharply reverse the consensus painfully developed over two hundred years of American history established themselves so powerfully in a scant ten years? Where do they find their support? This support is not extensive, and many interests and groups stand against these policies— as do the principles developed over a long period of time, written into the Constitution and the laws, and accepted by the great majority of the people and their political representatives. How do measures which most people oppose become imposed upon them in a democracy? There is one simple answer, with a good deal of truth in it. The people do not determine what is constitutional; the courts do. Whatever the people would have preferred to do, the courts insist on the Constitution. This may explain busing for school desegregation, but it will not satisfy us completely. The Federal courts, important as they are, do not explain everything. The Executive branch of government plays a great role, particularly in the area of employment. Even the most popular branch, the Congress, bears substantial responsibility for these policies. In 1972, when there was already opposition on the part of business, labor, and the academic community to the emerging pattern of hard affirmative action, Congress voted additional powers to the Equal Employment Opportunity Commission. Congress well knew these powers would be used to follow up immediately on the EEOC's victory over the largest private employer in the country, and to put pressures on states and cities every where to follow patterns of proportional representation in employment. Clearly Congress could have—one would think it would have—accepted the point of view of such major constituencies as business, labor, the universities, and city and state governments, most of whom would have been happy to have been relieved of the burden of recordkeeping, test changing, defense, litigation, and the like that the Congress was imposing on them.

Why, then does the new course of hard affirmative action have strength? The protected minority constituencies, the direct beneficiaries of measures of strong affirmative action in employment, form only 16 or 17 percent of the population of the country and a considerably smaller proportion of its voters, owing to an age distribution among them skewed toward younger ages, as well as a lower percentage of registration and voting for blacks in many Northern jurisdictions and Mexican Americans and Puerto Ricans in others.

Any simple interpretation in terms of the distribution of power certainly does not work. We must add another factor, the distribution of morality, or rightness, as it is felt by the best-educated and most enlightened parts of the community, and by a considerable part, too, of the affected electorate. The Mayor of Boston has never come out against busing, pure and simple; yet he has won two elections for mayor against Louise Day Hicks, who has done so, in a city in which the overwhelming part of the electorate is white and opposed to busing. One reason he won is that, in these earlier elections, busing seemed remote. He might well lose today. Another reason he won is that there are always other issues than busing in an election, and on these he has made a greater appeal than his opponent. Racial issues are always in competition with other, primarily economic issues, and the voters who would be against the liberal candidate because of his stand on racial issues may support him because of his stand on economic issues. The weight of these in the voter's mind will change, depending on the state of the economy and the nearness of a court-ordered busing requirement. Another factor reducing the weight of busing in voting generally is that busing is never decreed nationally or statewide. It hits one community after another, this time Pontiac, that time Detroit, next Boston. In the individual community, the issue is one of great moment. For the state it is not, for the nation it is not, for the impact of busing is spotty, and as long as it is, the voter believes that the community affected may well deserve the harsh measures that have been decreed for it. Thus he can vote both his sense of justice or morality at the same time that he does not oppose his self-interest.

Sometimes busing is a statewide issue, as it is or has been in the South where many communities have been required to bus simultaneously (Florida in the primary elections of 1972), or in Michigan, where a major court decision affected not only Detroit but its surrounding suburbs and therefore a good part of the population of the state. This decision was not only of statewide significance in the Democratic primary of 1972, but it affected the actions of the House of Representatives in trying to stop busing in 1971 and later years.

There are, I believe, three separate elements to the moral authority of hard affirmative action in the areas of employment and school integration.

First and preeminently, there is the moral authority given by the fact the Negro was enslaved and then held in subordination, by law in a good part of the country, by accepted custom in a good part of the rest. A policy that is proposed as being for the benefit of the black, for the redress of his situation, for the correction of a monumental wrong, carries enormous moral authority. This authority means that the individual action proposed, unless one is personally closely affected, is not really closely examined; it is assumed that what is proposed is right and can only be opposed from an inferior moral position or an actually immoral one.

The situation is similar to that of the case of equality: "In Western civilization at least, men have always believed that equality is in some sense the norm from which inequality represents a deviation," Irving Kristol writes. Or, to quote Richard Wollheim and Isaiah Berlin, "If I have a cake and there are ten persons among whom I wish to divide it, then if I give one tenth to each, this will not, at any rate automatically, call for justification; whereas if I depart from this principle of equal division I am expected to produce a special reason." [1]

It is in this sense that any claim for blacks (and in lesser degree for other minorities) has an immediate moral force and justification. The argument then *against* the extension of power to the EEOC or the even division of the children by race in the schools needs special justification. It needs to resort to States' Rights or to a general principle of freedom (which today has lesser weight than the opposing principle of equality), or to the discussion of matters in detail—as to just *what* does the EEOC decree for an individual company, on *what* evidence, or *what* is the specific impact of such a distribution of schoolchildren on the children, or their parents, or the city. Such concrete discussion inevitably leads to some impatience when contrasted with the grand principle of equality, impatience at least among those who are not directly affected.

This major pillar of morality is supported by two others. The first is the general view that the actions of the Supreme Court and the lesser Federal courts in determining what is the law and the Constitution not only demand assent on pragmatic grounds—because, after all, that is the way disputed issues on the meaning of the Constitution and the laws and the legitimacy of public and private action get settled—but must also be supported on moral grounds. It is necessary for all authorities not only to agree that these determinations are the law but to give them the fullest measure of political and moral support. Pragmatically, there is the legitimate fear that disorder and anarchy will prevail if the means by which complicated issues are settled in a complex society are themselves brought under attack; thus in these disputed areas, the fullest measure of support from all sources—political and moral as well as legal—is necessary if order is to prevail. The other side of the moral approval that is associated with the decisions of the Federal courts is moral disapproval of those who attempt to reverse the action of the Federal courts, even by the perfectly legitimate means of new legislation or constitutional amendment. The processes of legislation and amendment of the Constitution are long and complicated. They must be associated with a great deal of open argument and conflict to mobilize opinion and support. If one part of the population attempts to organize to amend the Constitution so as to change a certain line of decisions by the Federal courts, this leaves open and unsettled the question the courts have tried to settle. Thus an aura of moral disapproval hangs over efforts to challenge the Supreme Court on such matters as school prayers, or abortion, or busing.

This disapproval is communicated by the mass media, which today largely reflect liberal and educated opinion. (Educated opinion means, in this country—statistically speaking—liberal opinion.) Educated and liberal opinion has not always been so fervent and steadfast in its support of Supreme Court decisions. Violently opposed to the conservative court of the 1930s, it began to change rapidly as the court changed to accept a substantial measure of governmental intervention in economic matters, to become a strong defender of expanded civil liberties, and most important, to take up the leading position in striking down the laws and practices which segregated and subordinated the Negroes. By now there is an automatic assumption that morality demands that the decisions of the Federal courts be defended. Whether liberal and educated position would continue to support the courts if they began to change with new appointments of traditional liberals or conservatives is another question.

The impact of this massive insistence of the past twenty years that the decisions of the court are moral may be seen reflected in ordinary people. One typical kind of mass media event in recent years has been interviews with people as they prepare to send their children off to distant schools of which they are doubtful. They will often say to questioning reporters from the press or television, "I don't like it, but the court has ordered it and we must obey and I suppose it will be for the best in the end." The mass media have almost universally tended to support such a response not only because they are dominantly liberal in orientation and approve the decisions of the court, but because they generally feel a responsibility to support law and order. (Recently, as a result of the civil rights and student revolts of the 1960s, the mass media support of law and order has become more selective.[2])

A third element is involved in the moral authority of strong affirmative action: People who oppose racism do not want to find themselves in the posture of the South. When one reads through the lengthy Congressional debates on the expansion of the powers of the EEOC in 1972, one feels this is perhaps the strongest motivation for defending action that congressmen find in some sense wrong or unwise. The congressmen seem to be saying: How can we put ourselves in the position of those in the South who for so long resisted action which was just and necessary? The amendments proposed and sometimes adopted by Congress to protect against extreme action by the Federal government and the courts are often proposed by Southern representatives and smack of the measures that were used by the South in the 1950s and 1960s to resist school desegregation. There are excellent Northern legal authorities who look with doubt upon the enormous expansion of court powers in very complex matters formerly in the hands of state legislatures and school boards. But for many people, to be in the posture of supporting a measure the South once supported or still supports has the same effect as supporting a measure in the 1930s or 1950s that Communists supported.

Thus the measures we have attacked in Chapters 2 and 3, even if they could not get the support of Congress today on a straight vote, and would certainly not get the support of the majority of the American people if asked, have one enormous advantage: They are seen as moral, and a moral advantage in politics, being on the side of right, is worth a good deal. Those who believe that these measures can no longer be defended on such grounds, that they now reflect a bureaucratic expansion into realms of life that should properly be reserved to local government and the people, are either Southerners who are believed to speak only from sectional interests, or conservatives and traditional liberals who have little influence on the shaping of the minds of journalists and opinion-makers.

The moral advantage is, of course, not everything: It is as important to consider how the American political system works if we are to understand why measures opposed by most people nevertheless become the law. Howard Sherain has argued that Congress has never assented to goals and timetables in employment, either in the Civil Rights Act of 1964 or in the extension of its powers in 1972. Why then do we have goals and timetables? Because Executive authority is, in large measure, independent of Congress, and it would be extremely difficult for Congress, even if it acted and thought as one man—which it does not—to regulate Executive authority. It is, however, also extremely difficult, as we know, for the President to regulate Executive authority. President Nixon said in 1972 that he was against quotas, but his agents operated to impose them with ever greater force. He said even more strongly that he was against busing, and there, it is true, he was apparently effective in getting his agents to act with more circumspection. It is the courts which impose the largest measure of busing. Here the Federal bureaucracy has lagged behind. But the permanent bureaucracy which deals with these matters prefers the position of the courts to that of the Chief Executive.

So the Chief Executive can on some issues make his position (which is, in this case, also the position of Congress) felt, but on others, he is ineffective. Why? The long and short of it is that the Chief Executive can only hope to be effective on a *few* matters, and he chooses not to dissipate his energies on quotas in employment or measures to integrate the schools. It would give him little political advantage and make him many enemies. On any given issue, there is a group of interested parties for whom that issue is everything, and a group of interests for whom that issue is only one of many concerns. For civil rights groups, the questions of affirmative action and busing are everything—every effort will be made to achieve success. For those who oppose them, such questions are not everything. Business has other interests, labor has other interests, mayors and voters have other interests. A small group for which one issue is everything may overcome a large group for which the issue is only one among many. (In cities where busing is about to be instituted, the issue does become everything for both sides, at least for a while.)

A President would get little political advantage from fighting his bureaucracy not only because the truly committed would mobilize all their energies against him, but because they have great influence in the mass media, which will uniformly support the stronger civil rights position against the weaker one, owing to their general outlook on these matters. Thus, when a civil rights official resigns in protest against the Executive—this happened a number of times during the first Nixon Administration—the major news media uniformly handle it as a case of noble and unselfish men and women truly committed to justice committing an act of self-sacrifice against a politically minded Executive seeking to sell out the blacks and the minorities to gain the support of the most backward and reactionary elements. The given case may or may not support such an interpretation: Perhaps the official is really resigning to take a better position with a foundation or in private business, and perhaps the Executive is truly concerned with the justice of measures that disrupt communities for no observable advantage, or that place groups in a position of preference. But to find this out would take examination of a complicated situation and its background that the busy reporter does not have the time or perhaps the capacity for—a simple dependence on one's established prejudices in this matter is the norm, and that is how most of the stories will get written. Thus, the President who acts against the supporters of strong affirmative action in his bureaucracy will not only earn the fierce antipathy of a small but deeply committed group, but a milder disapproval from the broad mass of educated opinion. He will gain strong approval only in the South (and a conservative President, who might want to moderate strong affirmative action, probably already has the South) and among small circles of principled conservatives and liberals with little influence.

But then why is the President in the position of having to openly oppose his Federal bureaucracy if he should want to enforce the laws as he and Congress understand them? Why does the bureaucracy—the Equal Employment Opportunity Commission, the Office of Civil Rights of the Department of Health Education, and Welfare, the Civil Rights Division of the Department of Justice, the Office of Federal Contract Compliance of the Department of Labor—so strongly support a statistical pattern approach to defining and remedying discrimination against the will of Congress and the Executive? (We are assuming here the President and his appointed officials rather than the permanent bureaucracy *are* the Executive. Conflict among them is possible and occurs.) A few general points can be made.

First, it is characteristic for specialized civil rights agencies of government to be staffed by members of minority groups and advocates of the most far-reaching definitions of discrimination and most far-reaching measures against whatever is defined as discrimination. There are understandable reasons for this. Minority group members and civil rights activists would find these jobs most appealing. Politically, executives will often find it easiest to appoint blacks, Mexican Americans, and Puerto Ricans to such jobs: One gains credit with minority groups, the applicants are expected to be interested in the work of the agencies, and they may be knowledgeable about the problems they deal with. Thus, when Leon Panetta took over the Office of Civil Rights in 1969, of its staff of 278, 129 were black; 15 Spanish-surnamed Americans; 2 American Indian; 1 Asian American; and 131 other.[3] The pattern is similar in other agencies, limited by the need for some to employ large numbers of lawyers—in which case they are often young lawyers committed to an activist interpretation of the Constitution and the laws which will permit them to enter as many situations as possible and define remedies as far-reaching as possible.

These latter features are characteristic of most bureaucracies and administrative agencies. They try to expand the scope of their powers, and civil rights agencies are no exception. We can be sure that when the last discriminatory act is performed in this country the antidiscrimination agencies will have reached a size that is far greater than it is today and that there will be no observable tendency for these agencies to propose reductions in their staffs and budgets as discrimination declines. There is nothing special about that.

Finally, politically appointed administrators of these agencies, even when they are not chosen from minority groups or from civil rights activists and militants, will rapidly become strong defenders of their agencies and advocates of the expansion of its staff and powers. There is nothing exceptional about civil rights agencies in this regard, either, for one of the most striking characteristics of our politically appointed leaders of government agencies is that they are immediately and apparently willingly "captured" by their staffs, advocating the staff's view of its mission before Congress, defending the staff, demanding greater powers and more money for it. It would take us far afield to explain this phenomenon: One reason for it is that the politically appointed chief is dependent on his staff for information and arguments; a second, paradoxically, is that he does not expect to stay for long and it is better to be known when he leaves as someone who did a "good job" (that is, he got Congress to give him more money and staff, he got into the newspapers for running a live agency, and so on) than as someone who presided over the reduction or dismantling of an unnecessarily bloated office. In any case, the latter job, which will give him little credit and make him many enemies, is far harder than the former.

But aside from these common phenomena of the world of government bureaucracy, there are distinctive things about civil rights and civil rights agencies. Congress is often angry at these agencies for going beyond the powers Congress thought it had assigned. This should make these agencies more cautious. It does, but not as much as one might expect, because against the greater anger of Congress they have the greater protection of the courts. In any case, the civil rights activist has the moral advantage we have pointed to earlier—he knows he is right, he knows he is fighting for justice and against discrimination, he knows that every overreaching of his powers (in Congress's mind) is for a good and noble cause. He is given the strength of a relatively small and embattled fraternity, which has fought and worked together for decades, and knows its enemies are Congress (even when it passes civil rights laws) and the politically minded President (even when he enforces them). The attitude to the President is literally to the office, whoever the occupant. Kennedy was viewed as suspiciously by civil rights activists in government civil rights offices as Nixon (one heard in such offices, in the early 1960s, questions such as, why did he take until 1962 to fulfill his campaign promise to outlaw discrimination in government subsidized housing? And why didn't his order, when it was issued, cover housing built earlier with FHA insured mortgages?).

But if this is the case, why does not Congress intervene more strongly to define exactly what it means by discrimination and segregation, and what it feels are legitimate efforts to enforce it? We should, of course, point out that Congress includes not only the majority which passed amendments excluding ethnic imbalance as such in employment and school distribution from action by Federal agencies, but also substantial minorities (and, on occasion, majorities) deeply committed to the more expansive conceptions of discrimination and the granting of greater powers to combat what is defined as such. The civil rights agencies have their defenders as well as their critics. In any case, for Congress to act on detail is exceedingly cumbersome: If almost all of Congress is agreed, if the issue is technical or not controversial, it can act rapidly. In other cases, however, the arts of delay are as well-known to congressmen committed to the broad definition of discrimination and segregation as they are to Southerners trying to limit the powers of government agencies.

As Howard Sherain writes regarding what he calls "the illusion of Congressional acquiescence to Affirmative Action":

This argument [that Congress acquiesces] would run along these lines: "Congress knows of the use of Affirmative Action through contract compliance. Congress could have prohibited this use of Affirmative Action, had it desired. That [it has not] . . . indicates that Congress is in favor of it." This argument, however, is based on a misunderstanding of the workings of Congress. By ignoring the vast decentralization of power in Congress, it ignores the difficulty of bringing Congress to the point of action. This "congressional acquiescence" argument would lead us to believe, for instance, that the thirty-year existence of the Committee on Un-American Activities necessarily means Congress favors its continued existence. The truth of the matter, however, may well be that a majority of congressmen are not in favor of the Committee's existence, but do not consider the attempt to do away with the Committee as a wise use of their limited resources. It is similarly fallacious to conclude that simply because Congress has not objected to Affirmative Action, Congress therefore must be in favor of it.[4]

But the principal point is that the power of Congress, which may on occasion be exerted, is matched—and overmatched— by the power of the courts. Congress has clearly indicated its intent to exclude statistical disproportions in employment and in school distribution from the operations of the antidiscrimination laws. The Supreme Court and subordinate courts have ruled otherwise, and what they rule is the law of the land. There is no way for Congress to prevent busing to overcome uneven racial distributions in the schools. Whatever it does, and it has passed many amendments, the courts will coolly decree that what they have found is unconstitutional discrimination, and a finding of unconstitutional public action requires a remedy. Similarly, what Congress defines as a perfectly innocent racial or ethnic distribution of employees may be found by Federal courts to be the result of unconstitutional discrimination; once again, whether the discriminator is the corner delicatessen, or AT&T, or any state, or city, or the Federal government itself, remedies can be required. The only limit on those remedies are the Supreme Court (which itself set down the pattern under which such findings are made) or constitutional amendment. The introduction of statistical remedies in employment arises basically from the executive order and the agencies interpreting and carrying it out. The courts play a larger role in the imposition of statistical remedies for school desegregation. The incipient effort to introduce requirements for even racial and ethnic distribution in housing development is being pursued in the courts as well as the Executive. But all three efforts are ultimately backed up by judicial findings and rulings and are thus impregnable to congressional intervention.

So why do the courts rule the way they do? Why do they find the use of intelligence tests, high school diplomas, or arrest records discriminatory in selecting employees; why do they find schools that select academically superior students through examination discriminatory; why do they set aside union contracts and the complex pattern of employment-relations law to impose quotas; why do they supersede elected school bodies to unilaterally dictate expensive and disruptive patterns of student transportation? The short answer is that they have found unconstitutional discrimination and segregation and they have imposed the appropriate remedy. It is not an adequate answer. One reason it is not is that courts have never imposed and required such massive social change as they have in recent years. One must go back to the Supreme Court of the Thirties —which decreed the National Recovery Act and the Agricultural Adjustment Act unconstitutional—to find such sweeping interventions in government, setting aside the acts of elected bodies and officials and replacing them with the judgment of courts. But even these actions of the 1930s were not comparable to what we see today. They were sweeping in decreeing what government could *not* do, but not sweeping in imposing on government and business through judicial decree tasks that were difficult, expensive, disruptive, and very likely unachievable for all but short periods. The legal analysts who disagree with the course that the courts have taken find errors of analysis which have permitted such astonishing developments, and undoubtedly, this is one explanation. I believe the explanation lies in another sphere: in the large cultural and intellectual changes that have affected those who become lawyers, write briefs, and analyze them for Federal judges.

The judges follow the weight of judicial analysis and opinion. They are affected by it because, in these complex cases, they must be guided by what is set before them in lengthy briefs and analyses of complex facts. These analyses are then digested by other lawyers for them, generally the brightest graduates of the law schools. But this entire process is guided by the weight of educated opinion, an educated opinion which is convinced that morality and progress lie on the side of the broadest possible measures of intervention to equalize the employment of members of minority groups in every sphere and level of employment, and to evenly distribute students and teachers and administrators through school districts, expanded as broadly as possible to make this distribution approximate a national statistical norm. We are now engaged in a process whereby the arguments that have legitimated statistical disproportions as evidence of discrimination and segregation and statistical remedies as the proper ones to overcome the conditions found are now being applied to residential distribution. How is this educated opinion made? It is not easy to give answers, but we know what the dominant opinion is. We know, for example, that it is far easier to find experts to testify before courts that statistical disproportions are a proof of discrimination or segregation than to find experts to testify to the opposite: understandably so, because these are the dominant opinions of sociologists and demographers. Now a constitutional principle should be able to stand against such findings, but that is not the way judges work for the most part these days. Constitutional argument can be found to defend any principle, apparently, so the issue becomes not what is the principle but what is good for society.

Here, facts are assumed—they are the facts that are presented in textbooks, presented by professors and journalists in the mass media, and believed by lawyers who are, after all, coming through the universities in which these facts are developed and diffused—that are not true, but serve as the basis to guide judicial decision. Thus, it is believed that minorities cannot make progress even if discrimination in employment is effectively outlawed, and that they can only make progress if quotas are set for them. As we have seen, this is not the case. Blacks made progress as discrimination declined in the 1960s under the impact of law and changing attitudes. It is believed that any concentrations of black students and teachers in schools must be due to segregation. As we have seen, that is not true either. It is believed that blacks cannot learn unless they are in schools in which they approximate their proportion in some larger population. This outlandish proposition is of course untrue. It is believed that no blacks would live near other blacks if they could, but would distribute themselves evenly through a larger population. This is most unlikely. It is believed that resistance to quotas and school busing comes from racism, from resistance to the implementation of constitutional rights. Which is also largely not true. And all those who believe these things believe that they are moral and the others who oppose them are immoral, that they defend the best traditions of the United States, whereas those who disagree with them would perpetuate discrimination, segregation, and inequality.

Thus we have a complex of education, culture, law, administration, and political institutions which has deflected us into a course in which we publicly establish ethnic and racial categories for differential treatment, and believe that by so doing we are establishing a just and good society. Behind it all stands, to my mind, a radical misunderstanding of how we in the United States have attempted to deal with the problems of a multiracial and multiethnic society. The pattern we have developed is not easily summed up in slogans—which is perhaps its defect—for we have decided against both the forcible assimilation of all groups into one mold and the legal recognition of each group for the establishment of a formal parity between them. It is a pattern that has emerged from the complex interplay of constitutional principles, political institutions, American culture, and that has had, at times, to be reestablished through force and violence.

For ten years now, we have drifted in another direction, certainly in some ways an easier one to understand, and in some ways even easier to institute: let us number and divide up (some of) the people into their appropriate racial and ethnic groups, and let equality prevail between them and the "others." But this has meant that we abandon the first principle of a liberal society, that the individual and the individual's interests and good and welfare are the test of a good society, for we now attach benefits and penalties to individuals simply on the basis of their race, color, and national origin. The implications of the new course are an increasing consciousness of the significance of group membership, an increasing divisiveness on the basis of race, color, and national origin, and a spreading resentment among the disfavored groups against the favored groups. If the individual is the measure, however, our public concern is with the individual's capacity to work out an individual fate by means of education, work, and self-realization in the various spheres of life. Then how the figures add up on the basis of whatever measures of group we use may be interesting, but should be no concern of public policy.

This, I believe, is what was intended by the Constitution and

the Civil Rights Act, and what most of the American people—
in all the various ethnic and racial groups that make it up—
believe to be the measure of a good society. It is now our task
to work with the intellectual, judicial, and political institutions
of the country to reestablish the simple and clear understanding that rights attach to the individual, not the group, and that
public policy must be exercised without distinction of race,
color, or national origin.

NOTES

CHAPTER 1

1. Yehoshua Arieli, *Individualism and Nationalism in American Ideology,* Cambridge, Massachusetts: Harvard University Press, 1964 (Penguin Books Edition, 1966, pp. 19–20).

2. Arieli, *op. cit.,* pp. 86–87.

3. Seymour Martin Lipset, *The First New Nation,* New York: Basic Books, 1963, esp. Chap. 2, "Formulating a National Identity," and Chap. 3, "A Changing American Character?"

4. Hans Kohn, *American Nationalism: An Interpretive Essay,* New York: Macmillan, 1957 (Collier Books edition, 1961, p. 144).

5. Arieli, *op. cit.,* p. 44.

6. Kohn, *op. cit.,* pp. 143–144, quoting from Max Savelle, *Seeds of Liberty: The Genesis of the American Mind,* New York: Knopf, 1948, p. 567 f.

7. Kohn, *op. cit.,* p. 143.

8. On the legal position of Jews, see Oscar and Mary F. Handlin, "The Acquisition of Political and Social Rights by the Jews in the United States," *American Jewish Year Book,* Vol. 56, New York: American Jewish Committee, 1955, and Philadelphia: The Jewish Publication Society of America, 1955, pp. 43–98. There is no equally convenient summary for the legal position of Catholics, but see Anson Phelps Stokes, *Church and States in the United States,* New York: Harper and Bros., 1950, Vol. I, Chapters V and XII.

9. Arieli, *op. cit.,* pp. 27–28.

10. From *Redburn: His First Voyage,* quoted in Kohn, *op. cit.,* pp. 153–154.

11. From *The American Democrat,* New York: Knopf, p. 135, quoted in Kohn, *op. cit.,* p. 162.

12. Arieli, *op. cit.,* pp. 293–305.

13. Arieli, *op. cit.,* p. 292.

14. John Higham, *Strangers in the Land: Patterns of American Nativism, 1860–1925,* New Brunswick, New Jersey: Rutgers University Press, 1955, pp. 20–21.

15. For this period, see Higham, *op. cit.;* Seymour Martin Lipset and Earl Raab, *The Politics of Unreason: Right-Wing Extremism in America, 1790–1970,* Chaps. 3–4, New York: Harper & Row, 1970; C. Vann Woodward, *The Strange Career of Jim Crow,* New York: Oxford University Press, 1955.

16. Kohn, *op. cit.*, p. 161.

17. Kohn, *op. cit.*, p. 143.

18. Barbara Miller Solomon, *Ancestors and Immigrants: A Changing New England Tradition,* Cambridge, Massachusetts: Harvard University Press, 1956, p. 176.

19. Horace M. Kallen, *Culture and Democracy in the United States,* New York: Boni and Liveright, 1924, p. 93. Kallen does not give the writer's name but describes him as "a great American man of letters, who has better than anyone else I know interpreted to the world the spirit of America as New England." The writer was probably his teacher at Harvard, to whom he dedicated this book, Barrett Wendell.

20. Henry James, *American Notes,* New York: Charles Scribner's Sons, 1946 [originally 1907], p. 86.

21. Seymour Martin Lipset analyzes the significance of the difference in the political origins of the United States and Canada in "Revolution and Counterrevolution: The United States and Canada," Chapter 2 in *Revolution and Counterrevolution: Change and Persistence in Social Structures,* New York: Basic Books, 1968.

22. Marcus Hansen, *The Immigrant in American History,* Cambridge, Massachusetts: Harvard University Press, 1940, p. 132.

23. *Pierce* v. *Society of Sisters,* 268 U.S. 510 (1925). Of course, as public and bureaucratic controls multiply in every part of life, this freedom is restricted, and not only for ethnic groups. It means that the establishment of church or school in a single-family house—a typical pattern—may run into zoning and planning restrictions, and often does; that the establishment of a nursery or an old-age home in less than institutional quarters fulfilling state requirements becomes almost impossible. New groups suffer probably more from these restrictions than old groups. But I note that a magnificent nineteenth-century Richard Morris Hunt-designed home for the aged in New York, maintained by an old Protestant welfare agency, is to be demolished because it cannot meet state standards of "proper" facilities for the aged. All suffer from the ever-widening reach of state controls.

CHAPTER 2

1. *Equal Educational Opportunities Act, Hearings before the Committee on Education and Labor,* House of Representatives, 92nd Congress, 2nd Session, 1972, pp. 1408–1409.

2. *Legislative History of the Equal Employment Opportunity Act of 1972. . . .* Prepared by the Subcommittee on Labor of the Committee on Labor and Public Welfare, United States Senate, 92nd Congress, 2nd Session, November 1972, pp. 209–211.

3. *New York Times,* "U.S. Push for Job Equality," October 21, 1973, Section 3; *New York Times,* "New Powers Urged for Job Rights Agency," March 25, 1975.

4. "Integration Case of 1954 Recalled," *New York Times*, March 24, 1974.

5. *Boston Globe*, March 24, 1974.

6. Ben J. Wattenberg and Richard M. Scammon, "Black Progress and Liberal Rhetoric," *Commentary*, 55:4, April 1973, pp. 35–44; and letters on it in *Commentary*, 56:2, August 1973, pp. 4–22.

7. *The Social and Economic Status of the Black Population in the United States, 1973*, U.S. Bureau of the Census, Current Population Reports, Special Studies, Series P-23, No. 48, pp. 24, 25, 73, 54, 55.

8. Richard B. Freeman, "Changes in the Labor Market for Black Americans, 1948–1972," *Brookings Papers on Economic Activity*, 1973: 1, p. 118.

9. Robert E. Hall and Richard A. Kasten, "The Relative Occupational Success of Blacks and Whites," *Brookings Papers on Economic Activity*, 1973:3, pp. 785, 791–792.

10. Freeman, *op. cit.*, p. 67.

11. Elliott Abrams, "Employment Quotas: Issues in Law and Social Policy," 1973, unpublished.

12. *Harvard Law Review*, 84:5, March 1971, pp. 1127, 1128, 1130.

13. "Constitutional Requirements for Standardized Ability Tests Used in Education," *Vanderbilt Law Review*, 26:4, May 1973, footnote 122, p. 819.

14. Elliott Abrams, "The Quota Commission," *Commentary*, 54:4, October 1972, p. 57.

15. United States Commission on Civil Rights, *The Federal Civil Rights Enforcement Effort—A Reassessment*, 1973, pp. 52–54.

16. United States Commission on Civil Rights, *Federal Civil Rights Enforcement Effort, 1970*, p. 83.

17. On February 27, 1975, the U.S. Circuit Court of Appeals for the District of Columbia ruled that the Federal Service Entrance Examination was unlawfully discriminatory. The basis of the decision was a statistical study by the Urban Institute which showed that graduates from colleges with 99 percent black enrollment did much more poorly in the exam than graduates from colleges with 99 percent white enrollment. Owing to this finding, it was incumbent on the Civil Service Commission to show that the FSEE was "valid" (see p. 52 above), and the court ruled that it failed to do so. (*Search*, A Report from the Urban Institute, Vol. 5, Nos. 1–2, Spring 1975, p. 2)

18. This quotation, and those that follow on the guidelines, are from an extremely informative article by Michael J. Malbin, "Employment Report/Proposed Federal Guidelines on Hiring Could Have Far-reaching Impact," *National Journal*, September 29, 1973, pp. 1429–1434.

19. Malbin, *op. cit.*, p. 1432.

20. Equal Employment Opportunity Commission, *7th Annual Report* (undated, but for fiscal 1972), p. 11.

Notes

21. Equal Employment Opportunity Coordinating Council, Discussion Draft, August 23, 1973, "Uniform Guidelines on Employee Selection Procedures," p. 4.

22. Malbin, *op. cit.*, p. 1433.

23. Equal Employment Opportunity Commission, *7th Annual Report*, p. 11.

24. See, for something of this story, which is still not recorded in adequate detail, Paul Seabury, "HEW and the Universities," *Commentary*, February 1972, pp. 38–44; Daniel Seligman, "How 'Equal Opportunity' turned into Employment Quotas," *Fortune*, March 1973, pp. 160–168; a speech by Senator James Buckley on May 22, 1973, with a group of articles reprinted in the *Congressional Record* of that date; issues of *Measure*, published by the University Centers for Rational Alternatives, New York; and articles by Sidney Hook in *Freedom at Issue*, November–December 1971, March–April 1972, and July–August 1972.

On the legal aspects, the literature is voluminous, but *see* Howard Sherain, "The Questionable Legality of Affirmative Action," *Journal of Urban Law*, 51:25, 1973, pp. 25–47. Important reviews of the development of law are: George Cooper and Richard Sobol, "Seniority and Testing under Fair Employment Laws: A General Approach to Objective Criteria on Hiring and Promotion," *Harvard Law Review*, 82:8, June 1969, pp. 1598–1679; "Developments in the Law—Employment Discrimination and Title VII of the Civil Rights Act of 1964," *Harvard Law Review*, 84:5, March 1971, pp. 1109–1316. *See also* Stanley P. Hebert and Charles L. Reischel, "Title VII and the Multiple Approaches to Eliminating Employment Discrimination," *New York University Law Review*, 46:3, May 1971, pp. 449–485; Herbert Hill, "The New Judicial Perception of Employment Discrimination: Litigation under Title VII of the Civil Rights Act of 1964," *University of Colorado Law Review*, 43:3, March 1972, pp. 243–268.

Specifically on the issue of higher education, see *Higher Education Guidelines: Executive Order 11246*, U.S. Department of Health, Education and Welfare, Office of Civil Rights, October 1, 1972; and Richard A. Lester, *Antibias Regulation of Universities*, New York: McGraw-Hill, 1974; Sheila K. Johnson, "It's Action, but is it Affirmative?" *New York Times Magazine*, May 11, 1975, pp. 18 ff.

25. Sidney Hook, "The Road to a University 'Quota System,'" in *Freedom at Issue*, No. 12, March–April 1972, p. 21.

26. Barbara R. Lorch, "Reverse Discrimination in Sociology Departments: A Preliminary Report," *The American Sociologist*, 8:3, August 1973, p. 119.

27. *"A Unique Competence,"* A Study of Equal Employment Opportunity in the Bell System, EEOC Prehearing Analysis and Summary of Evidence before the FCC, Washington, D.C., Docket No. 19143, p. 240.

28. *Ibid.*, pp. 22–23.

29. *Federal Civil Rights Enforcement Effort, A Report of the United States Commission on Civil Rights,* 1970, pp. 70–71.

30. *Ibid.,* pp. 79, 80, 100.

31. See, on this report, Nathan Glazer, "A Breakdown in Civil Rights Enforcement?" in *The Public Interest,* No. 23, Spring 1971, pp. 110–111. The percentage of minorities in the Federal service has continued to increase since 1967. Minority group members accounted for 64 percent of all nonpostal federal employees hired between May 1973 and May 1974, despite the fact that there was a decline in total federal employment. Blacks held 14.6 percent of all nonpostal federal jobs in May 1974. Minority group members held 6.4 percent of GS-12 and 13 grade jobs, 5.1 percent of GS-14 and 15, 3.9 percent of GS-16 through 18. (*Washington Post,* June 12, 1975)

32. Speech by Senator James Buckley, *op. cit.*

33. The following table indicates the rate of change, the degree to which equality in proportion of high school graduates has been reached, and the remaining disparities:

Level of Schooling Completed by Persons 20 to 24 Years Old

	Male		Female	
	Black	White	Black	White
Percent completed 4 years of high school or more:				
1965	50	76	48	77
1973	70	85	72	85
Percent completed 1 year of college or more:				
1965	14	36	15	26
1973	27	46	25	37

Percent of Population 25 to 34 Years Old Who Completed 4 Years of College or More

	Black			White		
	Total	Male	Female	Total	Male	Female
1966	5.7	5.2	6.1	14.6	18.9	10.4
1973	8.3	8.0	8.5	19.0	22.6	15.5

Source: *The Social and Economic Status of the Black Population in the United States, 1973,* p. 69.

Notes

34. I have concentrated on the black population in this discussion because they are by far the largest minority—over 11 percent of the population. The next largest minority, as the Federal government lists minorities, the Mexican Americans, do not number even 3 percent of the population, the Puerto Rican not even 1 percent. I do not consider other "Spanish-surnamed" minorities necessarily deprived economically—the largest among them, the Cubans, have made good economic progress. "Orientals" are economically and educationally no worse off than the white population. From my own point of view, each group is so distinct that to deal with their problems in general would not make much sense. For this reason, I have concentrated on what is by far the largest group, blacks.

35. "Why Are New York's Workers Dropping Out?" in *Monthly Economic Letter,* August 1973, (New York: First National City Bank), pp. 12–13.

36. Martin Feldstein, "The Economics of the New Unemployment," *The Public Interest,* Fall 1973, pp. 3–42.

37. Bureau of the Census, *op. cit.,* Table 19.

38. Sar A. Levitan and Robert Taggart III, in *The Job Crisis for Black Youth,* Report of the Twentieth Century Fund Task Force on Employment Problems of Black Youth, New York: Praeger, 1971, p. 63.

39. Buckley, *op. cit.*

40. *Federal Register,* "Guidelines on Discrimination because of Religion or National Origin," January 19, 1973, Part (b) of 60–50.1.

CHAPTER 3

1. Gary Orfield, *The Reconstruction of Southern Education: The Schools and the 1964 Civil Rights Act,* New York: Wiley-Interscience, p. 1.

2. *Ibid.,* p. 348.

3. Leon Panetta and Peter Gall, *Bring Us Together: The Nixon Team and the Civil Rights Retreat,* Philadelphia and New York: J. B. Lippincott, 1971. For the hurricane story, *see* pp. 274–275.

4. In *Inequality in Education,* Center for Law and Education, Harvard University, August 3, 1971.

5. *Federal Civil Rights Enforcement Effort,* 1970, p. 14.

6. *Busing of Schoolchildren: Hearings before the Subcommittee on Constitutional Rights of the Committee on the Judiciary, United States Senate,* 93rd Congress, 2nd Session, 1974, p. 99.

7. Much will depend on changes in the Supreme Court. But *Swann* was a unanimous decision, and in *Keyes,* of four Nixon appointees, two voted with the majority, one concurred and dissented in part, and only one dissented.

8. For the question of the degree to which official action can be held responsible for residential segregation, *see* Chapter 4.

9. See Nathan Glazer, "Is 'Integration' Possible in the New York Schools?" in *Commentary*, 30:3, September 1960, pp. 185–193.

10. The *New York Times* reported on March 28, 1975:

School officials planning the integration of Coney Island's Mark Twain Junior High School have submitted a plan for extraordinary school-security measures, including ultrasonic detection devices and personal alarms for teachers, to assure the safety of the student body, which, under the integration plan, is to become predominantly white.

The security, beside the unparalleled concentration of electronic systems, also includes plans to take pupils home by bus in the middle of the school day in case of illness, transportation for parents to school meetings and a student-identification system.

Under measures apparently intended to eliminate the need or the opportunity for the students to wander around in the school's predominantly black and Hispanic neighborhood, the school has also proposed a compulsory in-school lunch program, immediate entrance into the school by students arriving in buses, and a fenced-in playground area with access through the school only.

The security proposals were submitted to Judge Jack B. Weinstein in Federal District Court as part of the integration order for the school that he handed down last July.

11. *New York Times,* December 11, 1973.

12. Lillian B. Rubin, *Busing and Backlash: White Against White in a California School District,* Berkeley, California: University of California Press, 1972, p. 163.

13. *Busing of Schoolchildren, op. cit.,* p. 100.

14. This has not kept "educational planners," intent on finding a means by which the single goal of reducing black students to a minority might be achieved, from proposing educational "complexes" ("parks") for 20,000 elementary schoolchildren. Even Federal judges have hesitated to order such solutions, though it is apparently within their power.

15. One may follow the story in the *New York Times,* March 2, 9; April 5, 25; May 23; July 3, 9, 11, 15, 22; September 22, 1973.

16. Derrick A. Bell, Jr., *Race, Racism and American Law,* Boston: Little, Brown, 1973, pp. 558–559.

17. Mexican American, Puerto Rican, and other Spanish-origin children as well as blacks are now regularly included in desegregation cases, and the required statistics include them. American Indians are, of course, counted separately. While Asian Americans have not complained of segregation, they, too, are counted and reported in the federally required school statistics. Some other groups—Portuguese Americans, French-origin children—are considered in some discussions as subject to segregation and statistics may shortly well be expanded to include them.

18. All the quotations on the situation in Charlotte are from the

Notes

testimony of William Poe, Chairman of the Charlotte-Mecklenburg School Board, as reported in *Busing of Schoolchildren,* pp. 95–98.

19. See John McAdams, "Can Open Enrollment Work?," *The Public Interest,* Fall, 1974, No. 37, pp. 69–88.

20. David K. Cohen, Thomas F. Pettigrew, and Robert S. Riley, "Race and the Outcomes of Schooling," in Frederick Mosteller and Daniel P. Moynihan, eds., *On Equality of Educational Opportunity,* New York: Random House, 1972.

21. In June 1975, an as yet unpublished study by James S. Coleman for the Urban Institute argued that there was an extensive withdrawal of white students from the public schools apparently as a result of busing, using statistics from the major school districts in the country. Thus, between 1968 and 1973 Atlanta public schools lost 51 percent of their white students, Memphis 43 percent. Of course part of this withdrawal could be attributed to the general decline of white population in central cities. But very likely requirements to attend distant public schools has had a substantial effect in speeding up this process. (*Newsweek,* June 23, 1975, p. 56)

22. Nancy H. St. John, *School Desegregation: Outcomes for Children,* New York: John Wiley, 1975.

23. See, for example, David J. Armor, "The Evidence on Busing," *The Public Interest,* No. 28, Summer 1972, pp. 90–126; Thomas F. Pettigrew, Marshall Smith, Elizabeth L. Useem, and Clarence Norman, "Busing: a Review of 'The Evidence,'"; David J. Armor, "The Double Standard: a Reply"; and James Q. Wilson, "On Pettigrew and Armor: an Afterword," *The Public Interest,* No. 30, Winter 1973, pp. 88–134.

24. *Equal Educational Opportunity: Hearings before the Select Committee on Equal Educational Opportunity of the United States Senate,* 92nd Congress, 1st Session, 1971, p. 4058 ff. For another, if less vivid, account of race relations in an integrated high school, see Bruce Porter, "It Was a Good School to Integrate," *New York Times Magazine,* February 9, 1975, on Madison High School in Brooklyn, which has been the subject of a study owing to racial clashes.

CHAPTER 4

1. These facts are drawn from a summary of census data in one of the most persuasive and thoughtful of the works arguing for a determined national effort to change the distribution of minority groups in metropolitan areas, Anthony Downs, *Opening Up the Suburbs: An Urban Strategy for America,* New Haven: Yale University Press, 1973, Appendix.

2. Downs, *op. cit.,* p. 189.

3. Bernard Frieden, "Blacks in Suburbia: The Myth of Better Opportunities," in *Minority Perspectives: Papers by Dale Rogers Marshall, Bernard Frieden, and Daniel William Fessler,* Washington, D.C.: Resources for the Future, 1973.

4. Similar, but more fragmentary, evidence is given in William W. Pendleton, "Blacks in Suburbs," in Louis H. Masotti and Jeffrey K. Hadden, eds., *The Urbanization of the Suburbs, Urban Affairs Annual Reviews*, Vol. VII, Beverly Hills, California: Sage, 1973.

5. *Metropolitan Bulletin*, published by the Washington Center for Metropolitan Studies, Washington, D.C., January 1974, Number 12; and the *Washington Post*, "Races Held Most Alike in Pr. George's," Section C, November 12, 1973.

6. *New York Times*, "City within a City Rising in Canarsie," March 21, 1974.

7. Iver Peterson, "Canarsie: Anatomy of a School Crisis," *Race Relations Reporter*, January 1973, pp. 9–12.

8. George Roniger, "Metro New York: An Economic Perspective," First National City Bank, Economics Department, 1974.

9. Karl and Alma Taeuber, *Negroes in Cities* (Chicago: Aldine, 1965.

10. It is an indication of how hard it is to find a general and satisfactory answer to the question of "how much segregation is there and what is the direction of change?" that the segregation indices from the 1970 census were not available and presented in a discussion paper of the Institute for Research on Poverty of the University of Wisconsin until February, 1974: Annemette Sørensen, Karl E. Taeuber, and Leslie J. Hollingsworth, Jr., *Indexes of Racial Residential Segregation for 109 Cities in the United States, 1940 to 1970*, Studies in Racial Segregation, No. 1.

11. Norman M. Bradburn, Seymour Sudman, and Galen L. Gockel, with the assistance of Joseph R. Neal, *Side By Side: Integrated Neighborhoods in America*, Chicago: Quadrangle, 1971, p. 11.

12. *Ibid.*, p. 12.

13. Angus Campbell, *White Attitudes Toward Black People*, Institute for Social Research, Ann Arbor, Michigan: University of Michigan, 1971, pp. 143, 144.

14. Frieden, *op. cit.*, pp. 42, 43.

15. U.S. Department of Commerce, *News*, Social and Economic Administration Statistics, March 12, 1974.

16. Bennett Harrison, *Urban Economic Development: Suburbanization, Minority Opportunity and the Condition of the Central City*, Washington, D.C.: The Urban Institute, 1974.

17. These rather startling reversals of widely accepted axioms are based on the research, principally, of Benjamin I. Cohen, "Trends in Negro Employment within Large Metropolitan Areas," *Public Policy*, Fall, 1971, Charlotte Fremon, unpublished papers for The Urban Institute, 1970; and Wilfred Lewis, Jr., unpublished manuscript, The Brookings Institution, 1969.

18. Frieden, *op. cit.*, p. 37.

19. See Chapter 2, pp. 42–43.

20. United States Commission on Civil Rights, *The Federal Civil*

Notes

Rights Enforcement Effort—1974, Volume II, To Provide . . . for Fair Housing, Washington, D.C., 1974, pp. 31, 52, 65, 126, 14, for the figures in text and in general for a full and detailed description of Federal efforts in this area.

21. See, for example, Anthony H. Pascal, "The Analysis of Residential Segregation," RAND Corporation, 1969, P–4234; and Karl E. Taeuber, "Patterns of Negro-White Residential Segregation," RAND Corporation, 1970, P–4288.

22. Nathan Kantrowitz, *Ethnic and Racial Segregation in the New York Metropolis: Residential Patterns among White Ethnic Groups, Blacks, and Puerto Ricans*, New York: Praeger, 1973, pp. 24–25 and Table 2.5.

23. Thomas C. Schelling, "On the Ecology of Micromotives," *The Public Interest*, Fall 1971, p. 82ff.

24. Pettigrew, *op. cit.*, p. 45.

25. *The Federal Civil Rights Enforcement Effort—1974, Volume II*, *op. cit.*, p. 85.

26. Different criteria could be used to assign multiple-family or small-lot housing to each community: Thus, Edward M. Bergman, in *Eliminating Exclusionary Zoning: Reconciling Workplace and Residence in Suburban Areas*, Cambridge, Massachusetts: Ballinger, 1974, proposes linking the amount of such housing to the number of jobs that could be created in the areas zoned for workplaces in a community. The community would have the responsibility of zoning so that the people who work there could live there.

27. The best and most convenient summary of the complex picture in this area is to be found in Charles M. Haar and Demetrius S. Iatridis, *Housing the Poor in Suburbia: Public Policy at the Grass Roots*, Cambridge, Mass.: Ballinger, 1974. For up-to-date information on the developing struggle in the courts to open up the suburbs, the Memoranda of the Metropolitan Housing Program of the Potomac Institute, Inc. (Washington, D.C.), are invaluable.

28. Jerome Pratter, "Dispersed Housing and Suburbia: Confrontation in Black Jack," in the American Society of Planning Officials, *Land-Use Controls Annual, 1971*, p. 151.

29. The Urban Institute, *Search*, March–April 1973.

CHAPTER 5

1. David Danzig, "Rightists, Racists and Separatists: A White Bloc in the Making?" *Commentary*, 38:2, August 1964, p. 28.

2. Danzig, *op. cit.*, p. 30.

3. S. M. Lipset and Earl Raab, *The Politics of Unreason*, New York: Harper & Row, 1970.

4. See Kevin Phillips, *The Emerging Republican Majority*, New Rochelle, New York: Arlington House, 1969.

5. See Richard Scammon and Ben Wattenberg, *The Real Majority,* New York, Coward, McCann & Geoghegan, 1970.

6. Common usage itself, we may point out, has undergone some surprising shifts. In the late 1960s, the Ford Foundation was reported to have a program of support of research by members of ethnic groups. Delighted, I wrote in for information, and was told that in the Ford Foundation's usage "ethnic" (then) meant members of racial minorities and of groups that had their origin in some parts of Latin America. Thus at that time, one usage of "ethnic" *excluded* all white ethnic groups except for some of Latin American origin. There is also an exactly *opposite* and more common contemporary usage in which "ethnic" is used to refer *only* to white groups, and to exclude blacks and Spanish-surnamed Americans. There is a vague common usage in which it seems to include everything that is not white Protestant and Anglo-Saxon, as in the term "ethnic" restaurants. For purposes of discussion by social scientists, however, ethnic has a fairly well-accepted meaning—a group with some degree of common cultural traditions and usages, defined primarily by descent, real or assumed. For the social scientist, the term "ethnic group" covers European white groups, racial groups, Puerto Rican and Mexican groups, and old Americans, who are also, after all, defined by some common cultural traditions, usages, and common descent, real or assumed. On "ethnicity" and the definition of "ethnic," see Nathan Glazer and Daniel P. Moynihan, eds., *Ethnicity,* Cambridge, Massachusetts: Harvard University Press, 1975, Introduction.

7. John Porter, "Ethnic Pluralism in Canadian Perspective," in *Ethnicity,* Nathan Glazer and Daniel P. Moynihan, eds., Cambridge: Harvard University Press, 1975, pp. 267–304; Wsevolod W. Issajiw, "Definitions of Ethnicity," *Ethnicity,* No. 1, July 1974, pp. 111–124.

8. U.S. Bureau of the Census, *Current Population Reports,* P–20, No. 249, "Characteristics of the Population by Ethnic Origin: March 1972 and 1971," and No. 250, "Persons of Spanish Origin in the United States: March 1971 and 1972," Washington, D.C.: U.S. Government Printing Office, 1973; and *Statistical Abstract of the United States, 1972,* p. 44.

9. Harold J. Abramson, *Ethnic Diversity in Catholic America,* New York: Wiley-Interscience, 1973, pp. 14, 19, 29, 31, 34, 39, 41.

10. Nathan Glazer and Daniel P. Moynihan, *Beyond the Melting Pot,* 2d ed., Cambridge, Massachusetts: M.I.T. Press, 1970, pp. xxxiii–xxxvi.

11. U.S. Bureau of the Census, "Characteristics of the Population by Ethnic Origin: March 1971 and 1972."

12. Edward O. Laumann, *Bonds of Pluralism: The Form and Substance of Urban Networks,* New York: John Wiley, 1973, p. 55.

13. Andrew Greeley, "Making It in America: Ethnic Groups and Social Status," *Social Policy,* September/October 1973, p. 29.

14. See Michael Novak, *The Rise of the Unmeltable Ethnics,* New

Notes

York: Macmillan, 1972; and review by Nathan Glazer, *National Review,* August 18, 1972, Vol. XXIV, No. 32, pp. 903–904.

15. Robert Kotlowitz, *Somewhere Else,* Charterhouse, 1972; Francine P. Prose, *Judah the Wise,* New York: Atheneum, 1973.

16. Andrew Greeley, "Political Attitudes among White Ethnics," *Public Opinion Quarterly,* 36:2, Summer 1972, pp. 216–217.

17. J. Michael Ross, "Race Issues and American Electoral Politics," paper presented to the American Political Science Association, September 1973, p. 28.

18. On the failure of blacks to open small business, compared with other groups, *see* Ivan Light, *Ethnic Enterprise in America,* Berkeley, California: University of California Press, 1972.

19. Gerald D. Suttles, *The Social Order of the Slum,* Chicago: University of Chicago Press, 1968, pp. 9, 47, 50–51, 52–53.

20. William Kornblum, *Blue-Collar Community,* Chicago: University of Chicago Press, 1974, pp. 50–51, 49.

21. The differences in the degree to which groups maintain family relationships by means of physical proximity is great. Thus, a NORC question on whether one's close relatives live in the neighborhood shows Italian and Polish far in the lead, every other group far behind. (Andrew Greeley, *Why Can't They Be Like Us?,* New York: Dutton, 1971, p. 77.)

22. From "EMPAC! Ethnic Millions Political Action Committee: To Build a New America," newsletter published by EMPAC!, P.O. Box 48, Bayville, New York, 11709.

23. Gary Jacobson, "Race by Any Other Name," in *Social Policy,* Vol. 4, No. 1, July/August 1973, pp. 40–41.

CHAPTER 6

1. Irving Kristol, "Equality As An Ideal," in *International Encyclopedia of the Social Sciences, Vol. 5,* New York: Macmillan, 1968, p. 108.

2. See, for example, Paul Weaver, "The New Journalism and the Old—Thoughts after Watergate," *The Public Interest,* Spring 1974, pp. 67–88.

3. Panetta, *Bring Us Together: The Nixon Team and the Civil Rights Retreat,* Philadelphia and New York: J. B. Lippincott, 1971, p. 128.

4. Howard Sherain, "The Questionable Legality of Affirmative Action," *Journal of Urban Law,* Vol. 51, No. 1, August 1973, p. 41.

INDEX

Ability tests, 36, 44, 51–57; differential validation of, 57–58; guidelines of Federal government on, 51–56; job performance and, relationship required between, 52–55, 57; for teachers, 65; validation of, 51, 52, 54–57, 65

Abrams, Elliott, 225n

Abramson, Harold J., 233n

Achievement: educational, see Educational achievement; as value, 11

Adams, John, 9

Affirmative action (in general): basic errors of, 202–204; community and fellow-feeling ignored by, 203; compensation for the past and, 201; confusion and inequalities with regard to composition of groups covered by, 198–200, 202; course of, in 1970s, 196–197, 220; educated opinion and, 209, 218–219; Federal courts's role in, 205, 208–211, 216–219; morality and, 205–211, 215; political system's and bureaucracy's role in, 211–219; resentment and hostility created by, 200–201; support for, sources and extent of, 204–206; two classes created by, 197–198, 203

Affirmative action in education, see Busing for racial balance; School desegregation

Affirmative action in employment (proportional representation requirements), 46–78, 187, 201, 203; arguments against, 69–76; colleges and universities and, 37, 38, 59–62; course of, in 1970s, 196–197; fixed ethnic-racial categories created by, 73–76; goals and timetables required for, 36, 37, 46, 48, 58, 59, 211; guidelines of Federal government on, 46–49, 56, 57, 74–75; imbalance and, 49; individuals as victims of discrimination and, 67–68; justification of, 68–70; quotas and, 34–37, 45, 49–52, 59, 60, 62, 64–66; tests for determining employee qualifications and, see Ability tests; "underutilization" criterion in, 47–49, 57, 66

Affirmative action in housing, 203; affirmative action in employment and education compared to, 134–135, 158–159; facts as to segregation and need for, 134–146; later white ethnic groups' resistance to, 187–189; local communities' powers and, 135–136; policies for, 151–167; practical problems raised by, 134–136; voluntary action as alternative to government-imposed, 165–167; zoning powers of local communities and, 135, 148, 161–165; see also Residential integration; Residential segregation

Affluence, assimilation and, 27–28

Index

Agnew, Spiro T., 80

American history: consensus of mid-1960s as culmination of, 5; direction or course of, with regard to ethnic questions, 5–32

American Indians, 6, 10, 17, 21, 28, 74, 201, 213, 229n

American Nationalism: An Interpretive Essay (Kohn), 9

"American" party (Know-Nothings), 15, 17

American Protective Association, 15, 17

American Revolution, 9–12, 16

Americanization, 184

Americans, self-naming as, 10–11

Anglo-Americans (Anglo-Saxon element), 12, 16–17, 173, 179

Anti-Catholic movement, 6, 17, 21

Anti-Chinese movement, 17, 21

Anti-immigrant movements, 6, 16–18; *see also* Exclusivism

Anti-Semitism, 17, 21, 182

Argentina, 30

Arieli, Yehoshua, 8–9, 12, 13, 223n

Arizona, 25

Armor, David J., 230n

Asian Americans (Orientals), 17, 74, 103, 124, 125, 199, 202, 213, 228n, 229n; *see also* Chinese; Japanese

Assimilation, 29–30; affluence and, 27–28; faith in nation's capacity for, 15–16; Jefferson's view of, 11–12

AT&T, 37, 62

Atlanta (Georgia), 107

Backlash among white ethnics, *see* Political reaction of white ethnic groups

Baltimore (Maryland), 141, 170

Belgium, 202

Beliefs, discrimination according to 193–194

Bell, Derrick A., Jr., 107–108, 229n

Bergman, Edward M., 232n

Berkeley (California): residential segregation indices for, 143; school desegregation in, 116, 124–126

Berlin, Isaiah, 207

Bilingual education, 104, 182–183

Bilingualism in Canada, 23

Biographical histories, as hiring criterion, 56

Black Jack (Missouri), 164–165

Blacks (Negroes), *see specific subjects*

Boston (Massachusetts), residential segregation indices for, 143

Boston Globe (newspaper), 40

Boston Latin high school, 92

Boston school system: desegregation (busing for racial balance) in, 104, 105, 107–109, 121, 206; entrance-by-examination high schools in, 92–93; quotas for hiring of black teachers in, 65–66

Bradburn, Norman M., 156, 231n

Bradley, Thomas, 147

Brooke, Edward W., 125, 126, 147

Brown v. *Board of Education*, 83, 95, 110, 127

Buckley, James, 226–228n

Buggs, John A., 40, 55

Bureaucracy, affirmative action and role of, 212–215

Busing for racial balance, 33, 40, 83–84, 86–90, 92, 101–103, 105, 107, 121, 124, 128, 140; in Charlotte-Mecklenburg district, 87–89, 112–114; educational achievement and, 122–123; elections and, 206–207; enrollment drops after court-imposed, 121–122; Gallup survey on attitudes towards, 84; later white ethnic groups' resistance to, 187; of low-achievement children, 118; in San Francisco, 90–92; voluntary, 116–119, 122, 124

Cahan, Abraham, 20
California, 101, 147
Cambridge (Massachusetts), 143
Campbell, Angus, 231*n*
Canada, 23–24, 173
Case, Clifford, 45
Caste system, 7, 8, 17, 22
Catholicism, 175, 179
Catholics, 6, 13, 16, 21, 123, 174, 175, 186, 223*n*
Census, U.S., ethnicity data of, 171–176, 179–180
Central cities: affirmative action for a more even racial and ethnic distribution between suburbs and, *see* Affirmative action in housing; black and white population trends in, 136–143; data on population, income, and jobs in, 132–133; effects of a more even redistribution of blacks between suburbs and, 146–151; houses built in, as a ratio to population growth, 132–133; jobs and income affected by black concentration in, 131–132, 149; residential-segregation indices for, 142–144
Central European immigrants, 16
Charlotte (North Carolina), 87–89, 104, 112–115
Chicago (Illinois): differences between black and white ethnic life styles in, 189–192; residential-segregation indices for, 143
Chicanos, *see* Mexican Americans
Chinese (Chinese Americans), 6, 17, 21, 41, 110, 183–184, 199
Churches, ethnic groups free to establish own, 25–26

Cincinnati (Ohio), 102
Cities, *see* Central cities; Ghetto; Metropolitan areas; Neighborhoods
Civil Rights, Office of, 116, 213
Civil Rights Act of 1964, 3, 4, 22, 35, 43–44, 50, 51, 61–62, 151–152; "affirmative actions" as used in, 58; individual rights, not group rights, granted by, 44–45; 1972 amendments of, 62, 64; provisions of, 44; school desegregation and, 78, 111; Title VII of, *see* Title VII of Civil Rights Act
Civil rights agencies, affirmative action and role of, 213–215
Civil Rights Commission (CRC), 39, 44, 54–55, 63, 82, 100, 159, 255*n*
Civil Service, Federal, 63–64
Civil Service Commission, 54–55, 57, 225*n*
Civil War, American, 14, 15
Clark, Joseph, 45
Clark, Ramsey, 93
Classes, social, 29; integration of different, educational effects, 120–121; school busing decisions and, 105; residential redistribution of blacks into areas of similar, 147–148
Cleveland (Ohio), 139
Cohen, Benjamin I., 150, 231*n*
Cohen, David K., 230*n*
Coleman, James S., 120–121, 230*n*
Coleman Report, 120–121
Colleges and universities, 6, 199; affirmative action in employment and, 37, 38, 59–62; black, Federal government recruitment in, 63–64
Commentary (magazine), 41
Commission on Civil Rights, U.S., 39, 44, 54–55, 63, 82, 100, 159, 255*n*

Index

Community control of (or influence on) schools, 104–106

Community feelings, affirmative action ignores, 203

Comte, Auguste, 14–15

Congregationalists, 180

Congress, affirmative action and role of, 205, 210, 211, 213–216

Consciousness, national (national identification), 8–9, 11, 15

Consensus on ethnic questions, 3–5, 203

Cooper, George, 226n

Cooper, James Fenimore, 14

Cornell University, 60

Councils of Government (COGs), 162

Courts, affirmative action and role of, 2–5, 208–211, 216–219

Covello, Leonard, 109

Crime, 188, 189

Criminal records, taken into account in hiring, as discriminatory, 56–57

Cubans, 74, 173, 198, 199, 228n

Cultural deprivation (cultural loss), sense of, 181–185

Cultural pluralism, 29, 30

Culture, American, 15

Danzig, David, 170, 232n

Democratic coalition, 169–171

Democratic party, 169

Democratic primaries of 1964, 169, 170

Denver (Colorado), 92

Desegregation: residential, see Affirmative action in housing; Residential integration; of schools, see School desegregation

Detroit (Michigan), 89, 99, 141, 206

Differential validity of ability tests, 57–58

Downs, Anthony, 150, 165, 166, 230n

Dutch, 12

East Chicago (Illinois), 143

East Cleveland (Ohio), 139

East Orange (New Jersey), 143

Eastern Europe, 22–23

Eastern European immigrants and their descendants, 6, 16, 18, 19, 21, 75, 170, 192; census data on, 173–176; cultural basis for political action among, 183–185

Economic inequality, 30–31

Education: bilingual, 104, 182–183; black-white differential expenditures on, effect of, 119–120, 146–147

Educational achievement: busing of children of low, 118; effects of school desegregation plans (racial balance) on, 119–124; employment and residential distribution related to, 130–131; expenditures and, relationship between, 147

Educational requirements for jobs, 56

Educational Testing Service, 55, 65

Edwards, Jack, 83

EEOC, see Equal Employment Opportunity Commission

Eisenhower, Dwight D., 46

Elections, 206

Elementary and Secondary Education Act (1965), 78

Eliot, Charles, 18–19

Emerson, Ralph Waldo, 10, 18, 19

EMPAC (Ethnic Millions for Political Action Committee), 192–193

Employment, 31; affirmative action in, see Affirmative action in employment; Civil Rights Act's provisions on, 44; enforcement of laws against discrimination in, 36–40; equal opportunity in, *see* Equal opportunity in employment; executive orders on discrimination and, 45–47; residence and education related to, 130–131

England, 10, 11

English origin, Americans of (Anglo-Americans), 12, 16–17, 173, 179

Equal employment opportunity, 44, 46–49; redefined as "result," 48–49

Equal Employment Opportunity Commission (EEOC), 205, 207, 210, 225*n*, 226*n*; affirmative action in employment and, 34, 37–40, 47, 49, 51–57, 62, 64, 67; tests for determining employee qualifications and, 51–57

Equal Employment Opportunity Council, 54, 55, 226*n*

Equal opportunity, as proper objective of public policy, 168–169

Equality: income, 41–42, 70; as value, 11, 207, 208

Equality of Educational Opportunity (Coleman), 120–121

Ethnic groups (ethnicity): ambiguous status of, 28–30; categorization of, confusions and inequalities in affirmative-action coverage, 198–200, 202; cultural deprivation or loss of, sense of, 181–185; deprivations of, 179–185; maintenance of institutions of, 25–29; as political (subnational) entities, limits on, 22–25; socio-economic inequalities related to growing attention paid to, 30–32; U.S. Census data on, 171–176; *see also specific groups and other subjects*

Ethnic identity, political action and, 176–179

Ethnic Millions for Political Action Committee (EMPAC), 192–193

European immigrants, *see* Central European immigrants; Eastern European immigrants and their descendants; Northern European immigrants and their descendants; Southern European immigrants and their descendants

Evanston (Illinois), 143

Examination high schools, 92–93

Exclusivism (opposition to immigration), 12–21

Executive branch (President), affirmative action and role of, 205, 211–213, 215, 217

Executive orders, 217; on employment discrimination, 45–47

Experience, black-white ethnic comparison of, 194–195

Fair Housing Act of 1968, 133, 151, 152

Fair Housing laws or ordinances, 133, 169–170

Families: decline in stability of black, 188–189; female-headed, *see* Female-headed families

Federal Contract Compliance, Office of (OFCC), 39, 49, 51

Federal courts, affirmative action and role of, 205, 208–211, 216–219

Federal Housing Administration (FHA), 93

Federal Service Entrance Examination (FSEE), 54, 55, 225*n*

Feldstein, Martin, 228*n*

Female-headed families: poverty among black, 71–72; rise in percentage of black, 41, 43

Index

FHA, 93
First National City Bank, 142
First New Nation, The (Lipset), 9
Florida, 206
Ford Foundation, 233*n*
France, 30
Franklin, Benjamin, 12, 17–18
Freedom, 207; as issue in school desegregation, 109–111, 118–119, 129
"Freedom of choice," as South's response to Federally-imposed school desegregation, 85, 118
Freeman, Richard B., 43, 225*n*
Freeman, Charlotte, 231*n*
French Americans, 173, 229*n*
French Canadians, 23–24
Frieden, Bernard, 139, 230*n*, 231*n*

Gall, Peter, 228*n*
Gardner, John, 60
Garrity (Judge), 65, 92, 93, 101, 105, 108, 109
German immigrants and their descendants, 12, 16–18, 173, 175, 179, 180, 186
Ghetto, 137; as a place of refuge, 106–107; suburbanization of blacks not an extension of, 139
Glazer, Nathan, 227*n*, 229*n*, 233*n*, 234*n*
Goals and timetables, affirmative action in employment and requirement for, 36, 37, 46, 48, 58, 59, 211
Gockel, Galen L., 231*n*
Goffman, Erving, 189
Greeks, 185
Greeley, Andrew, 180–181, 233*n*, 234*n*
Green, Edith, 34, 40
Griggs v. *Duke Power Company*, 52

Group identity, 5, 7
Guion, Robert M., 57

Haar, Charles M., 232*n*
Hall, Robert E., 43, 225*n*
Handlin, Mary F., 223*n*
Handlin, Oscar, 223*n*
Hansen, Marcus, 24–25, 27, 184, 224*n*
Harlan, John Marshall, 44
Harrison, Bennett, 149, 150, 231*n*
Harvard Law Review, 52, 53
Hawkins, Augustus, 34
Health, Education and Welfare, Department of (HEW): affirmative action in employment and, 37, 60–62; school desegregation and, 78–81, 85, 112
Herbert, Stanley P., 226*n*
Hesburgh, Rev. Theodore M., 39
Hicks, Louise Day, 170, 206
Higham, John, 15–16, 223*n*
Hill, Herbert, 226*n*
Hollingsworth, Leslie J., Jr., 231*n*
Hook, Sidney, 60, 226*n*
House Un-American Activities Committee, 216
Housing, 7, 31, 32; affirmative action in, *see* Affirmative action in housing; discrimination banned in renting or selling, 133, 151–152; effect of residential redistribution on quality of, 148–149; low-cost (low-income), Federally or state assisted, 158–162, 166; subsidized or assisted by government, 135, 158–163, 166
Housing Act of 1974, 160, 161
Housing and Urban Development, Department of, 151, 152, 160
Housing Appeals Committee (Massachusetts), 162
Houston (Texas), 141

Howells, William Dean, 20
Humphrey, Hubert H., 45

Iatridis, Demetrius S., 232*n*
Identification, national (national consciousness), 8–9, 11, 15
Identity: group, 5, 7; national, 8–9, 11, 15; political action and source of, 176–179
Immigrants: in Canada, 23–24; opposition to (exclusivism), 6, 12–21; *see also specific ethnic or nationality groups*
Immigration, 5; Cooper's views on, 14; Franklin's views on, 12; Melville's views on, 14; opposition to (exclusivism), 6, 12–21
Immigration Act of 1965, 3, 4
Income: affirmative action in housing and, 135; of blacks as proportion of white income, 42–43, 72; of central city v. suburban residents, 132; discrimination as factor in black-white differences in, 42–43; median, by ethnic group, 179–180; residence and educational achievement related to, 130–131
Income equality, of husband-wife black and white families, 41–42, 70
Income groups: low, Federally or state subsidized or assisted housing for, 158–162, 166; policies for producing an even residential distribution of, 135, 158, 163
India, 200
Indiana, 86
Indians, 199; American, 6, 10, 17, 21, 28, 74, 201, 213, 229*n*
Individualism and Nationalism in American Ideology (Arieli), 9
Inequality: economic, 30–31; social, 30–31

Inner cities, *see* Central cities
Institute for Social Research, University of Michigan, 145
"Institutional racism," 38, 68–69
Integration: of blocks and neighborhoods, data on, 144–146; ethnic and racial divisions created by affirmative action and, 202; Jefferson's views on, 11–12; of schools, *see* School desegregation; socioeconomic, educational effects of, 120–122; white ethnic groups' attitudes toward, 186
Intellectuals: New England, 16–19; sense of cultural loss among, 183
Interviews: of teacher applicants, in Boston, 65; unscored (unquantified), validity of, 55–56
Irish (Irish Americans), 19, 24–25, 179, 180, 186, 198, 203; census data on, 173–175
Israel, 185
Issajiw, Wsevolod W., 233*n*
Italians (Italian Americans), 19, 180–184, 203; census data on, 173–175; cultural deprivation of, sense of, 181–183; life-style differences between blacks and, 189–190, 192

Jacobson, Gary, 193–194, 234*n*
James, Henry, 20, 224*n*
James, William, 18
James v. *Valtierra,* 160
Japanese (Japanese Americans), 6, 21, 41, 183–184, 199
Jefferson, Thomas, 10–12
Jewish culture, 182, 183
Jews, 10, 13, 22, 23, 123, 157, 182, 198, 203, 223*n*; affluence and assimilation of, 27; census data on, 174–176, 179; cultural basis

Jews *(continued)*
for political action among, 183–185; as ethnic group, U.S. Census and, 172, 174; exclusivism directed against, 17, 21; *see also* Anti-Semitism
Jim Crow laws, 17
Jobs: central-city concentration of blacks and, 131–132; government, increase in, 150; residential redistribution of blacks and distribution of, 149–150; white ethnic-black conflict over, 186–187; *see also* Employment; Occupational distribution
Johnson, Lyndon B., 46
Johnson, Sheila K., 226n
Judaism, 175, 183
Justice, Department of: affirmative action in employment and, 51, 53, 54; school desegregation and, 78, 79, 112

Kallen, Horace M., 19, 224n
Kantrowitz, Nathan, 153–155, 232n
Karsten, Richard A., 43, 225n
Kennedy, John F., 46, 215
Keyes v. *School District No. 1, Denver,* 91–93, 98, 228n
Know-Nothings, 15, 17
Kohn, Hans, 9, 11–13, 223n, 224n
Kornblum, William, 191–192, 234n
Kotlowitz, Robert, 234n
Kristol, Irving, 207, 234n
Ku Klux Klan, 15, 21, 79

Labor, Department of, 37, 73–74
Labor-force participation of blacks, 71, 72
Labor unions: employment discrimination and, 50, 178; later white ethnic groups and, 186–187
Lake County (Indiana), 170

Language, sense of cultural loss and, 181–184
Latin Americans, *see* Spanish Americans
Laumann, Edward O., 233n
Lazarus, Emma, 18
Leftwich (associate superintendent of schools), 108
Legal Defense Fund, NAACP, 78
Levitan, Sar A., 228n
Lewis, Wilfred, Jr., 231n
Liberalism, 68, 93, 209
Liberals, 127, 128, 209, 210, 212
Life-style differences, between blacks and later white ethnic groups, 187–192
Light, Ivan, 234n
Lipset, Seymour M., 9, 223n, 224n, 232n
Lorch, Barbara R., 226n
Los Angeles (California), 143
Lowell, James Russell, 16, 18

McAdams, John, 230n
McGovern, George, 193
Madison, James, 9
Malamud, Bernard, 183
Malbin, Michael J., 225n, 226n
Marketing plan for housing, affirmative action, 159
Mass media, 209, 212
Massachusetts, 101; low-income housing in, 161–162
Mecklenburg County (Charlotte-Mecklenburg district) (North Carolina), 87–89, 112–115
Melville, Herman, 14
Metropolitan areas: residential-segregation indices for, by block, 142–144; *see also* Central Cities
Mexican Americans (Chicanos), 17, 21, 74, 171, 177, 191, 198, 201, 205, 213, 228n, 229n, 233n
Miami Valley Regional Planning Commission, 162

Michigan, 206–207
Mission Coalition, 104
Mississippi, 80
Mitchell, Clarence, 33–34
Mobility of American population, residential concentrations and, 155–157
Mondale, Walter F., 124
Morality (moral authority), affirmative action and, 205–211, 215
Mt. Vernon (New York), 143
Moynihan, Daniel P., 233n
Mulford, Elisha, 16

NAACP, 78, 100, 107
National identification (national identity or consciousness), 8–9, 11, 15
National Opinion Research Center (NORC), 144, 145, 186, 234n
National polities, limits set by inclusion process on formation of, 22–25
National Teachers Examination, 65
Nationalism, 15, 16
Nativism (nativist movements), 13, 15, 16, 170
Neal, Joseph R., 231n
Negroes in Cities (Taueber and Taueber), 142
Neighborhoods, 178; integrated, data on, 145–146; rapid change of racial or ethnic composition of, 155–157
New England intellectuals: immigration opposed by, 16–19; racism of, 16–17
New Jersey, 163
New Kent County (Virginia), 85–86
New Mexico, 25
New Rochelle (New York), 143

New York City, 104, 140–141, 170; labor-force participation of blacks in, 71; residential concentration of blacks in, 154–155; residential-segregation indices for, 143; Urban Development Corporation of, 161, 162
New York State, 101, 147
New York Times, The, 36, 38–40
Newark (New Jersey), 143
Nicaraguans, 198
Nixon, Richard M. (Nixon administration), 36, 79–80, 160, 211, 212
NORC (National Opinion Research Center), 144, 145, 186, 234n
Norfolk (Virginia), 121
Norman, Clarence, 230n
North, the: school desegregation in, 78, 81, 83, 89, 93, 95, 98; white backlash in, 169–171
Northern European immigrants and their descendants, 154, 155, 170
Novak, Michael, 183, 233–234n

Oakland (California), 143
Occupational distribution: changes in, for blacks, 70; by ethnic group, 175–176, 179; random, requirement for, 62–63; *see also* Jobs
Occupational identity, 176–178
OFCC (Office of Federal Contract Compliance), 39, 49, 51
O'Hara, James G., 33–34, 40
Orfield, Gary, 79–80, 228n
Orientals (Asian Americans), 17, 74, 103, 124, 125, 199, 202, 213, 228n, 229n; *see also* Chinese; Japanese

Paine, Thomas, 10
Panetta, Leon, 80, 213, 228n, 234n

Index

Pasadena (California), 121, 143

Pascal, Anthony H., 232*n*

Pendleton, William W., 231*n*

Pennsylvania, 12, 101

Peterson, Iver, 231*n*

Pettigrew, Thomas F., 230*n*, 232*n*

Philadelphia Plan, 34, 37, 50

Phillips, Kevin, 232*n*

Plessy v. *Ferguson,* 31, 44

Pluralism, cultural, 29, 30

Poe, William, 230*n*

Poles, 173–175, 179–185

Political action: cultural basis for, 181, 182, 184; identity and, 176–178

Political power of blacks, 128; residential redistribution and, 147

Political reaction of white ethnic groups (white backlash), 168–195; beliefs as basis of discrimination and, 193–194; comparison of experience and, 194–195; cultural loss or deprivation and, 181–185; to entry of blacks into their residential areas and schools, 187–189; jobs as source of white ethnic-black conflict and, 186–187; life-style differences between blacks and white ethnic groups and, 187–192; racism and prejudice and, 185–186, 193, 194

Political system, affirmative action and role of, 211–219

Polities, national, limits set by inclusion process on, 22–25

Population trends in central cities and suburbs, comparison of, 132–133

Populist movement, 17

Porter, Bruce, 230*n*

Porter, John, 233*n*

Portuguese Americans, 229*n*

Pottinger, J. Stanley, 80–81

Poverty: black families in, 71–72; in central cities, 141–142

Pratter, Jerome, 232*n*

Prejudice: ingrained, statistical approach justified by, 68; among later ethnic groups, 186; *see also* Racism

President, U.S. (Executive branch), affirmative action and role of, 205, 211–213, 215, 217

Prince Georges County (Maryland), 140, 141

Private schools, 105

Professional schools, 6

Prose, Francine P., 234*n*

Protestant religions, 12

Protestants: Anglo-Saxon, 179, 186; backlash by white, 170, 171, 186

Puerto Ricans, 74, 154, 155, 173, 177, 198, 205, 213, 228*n*, 229*n*, 233*n*

Quotas, for employment of minorities, 34–37, 45, 49–52, 59, 60, 62, 64–66

Raab, Earl, 223*n*, 232*n*

Race, Racism, and American Law (Bell), 107

Race relations, effect of residential redistribution on, 147–148

Race Relations Reporter, 140–141

Racial Isolation in the Public Schools (Coleman), 120

Racism, 19, 21, 93, 169, 209; as basis for political action among white ethnics, 185–186, 193, 194; as defining American history, 6–8; "institutional" or ingrained, 38, 68–69; of New England intellectuals, 16–17; *see also* Prejudice

Random distribution of minorities in jobs, 62–63
Reconstruction, 16, 17
Reconstruction of Southern Education, The (Orfield), 79
Recruitment: of black college graduates by Federal government, 63–64; of black teachers by Boston school system, 65–66
Reischel, Charles L., 226n
Religion, 10, 11, 15; *see also* Churches; *and specific religions*
Republican party, 15, 16, 171
Residence, employment and educational achievement related to, 130–131
Residential integration: benefits of, as ambiguous, 202; data on, 144–146; later white ethnic groups' resistance to, 187–189; stable, as difficult to achieve, 155–157
Residential segregation (residential concentrations of ethnic or racial groups), 130–167; affirmative action for overcoming, *see* Affirmative action in housing; cultural factors as cause of, 153; de facto vs. de jure, 93; discrimination as cause of, 153, 155; economic factors as cause of, 153, 166; enforcement of antidiscrimination laws and, 151–152; facts as to, 136–146; indices of, 142–143; mobility of American population and, 155–157; of Northern from Southern European groups compared to black-white segregation, 154–155; voluntary action to overcome; *see also specific topics under* Central cities; Suburbs
Restrictive covenants, residential selling or renting and, 133
Richmond (Virginia), 89

Richmond school district (California), 102
Riley, Robert S., 230n
Rokeach, Milton, 193
Roniger, George, 231n
Roosevelt, Franklin D., 45, 169, 170
Ross, J. Michael, 234n
Roth (Judge), 106, 109, 153
Rubin, Lillian B., 229n
Russia (pre-revolutionary), 29
Russian origin, Americans of, 173, 174, 179, 180

St. John, Nancy H., 122, 230n
St. Louis (Missouri), 164
San Francisco (California), 198; residential - segregation indices for, 143; school desegregation in, 90–92, 96, 99–101, 104, 109–111, 121
San Francisco State College, 60
Savelle, Max, 223n
Scammon, Richard M., 225n, 233n
Scandinavians, 180, 186
Schelling, Thomas C., 156, 232n
School boards: de facto and de jure segregation and obligations of, 100–102; voluntary action for racial balance and, 116–117
School desegregation (affirmative action for racial balance), 78–129, 202; advances made since 1954, 127–128; benefit of, as questionable, 201–202; busing for, *see* Busing for racial balance; Civil Rights Act provisions on, 78, 111; community-school relationship and, 103–106; confusion regarding groups covered by, 198, 199; course of, in 1970s, 196–197; court-imposed vs. voluntary, compared, 116–117, edu-

School desegregation *(continued)*
cational-achievement effect of, 119–124; enrollment drops after court-imposed, 121–122; freedom as issue in, 109–111, 118–119, 129; "freedom of choice" as South's response to Federal effort to impose, 85, 118; geographical zoning of school districts and, 85–92, 98–99, 102; guidelines for, 78–80; later white ethnic groups' resistance to, 187–188; merger of school districts for, 89, 90, 122; morality (moral authority) and, 207–208, 210, 211; relations between races as affected by, 123–127; in South, 78–89, 93, 95, 98, 118, 206; statistical-distribution test for, 50, 95–96; territorial or income classifications as substitute for racial classifications and, 117–118; transfer programs and, 88, 91, 92; transitional or temporary character of assignment of children by race or ethnic group to achieve, 111–115, 197; voluntary action to achieve, 116–119, 124; *see also* School segregation

School segregation: de facto vs. de jure, 89–93, 96–103, 115; educational effects of, 94–97; Federal funds denied to districts practicing, 78, 79; intent required for finding of de jure (unconstitutional), 91, 92, 98–99, 103; remedies as disproportional to scale of, 103–104; *see also* School desegregation

Schools, 6–7, 31–32; black-white differential expenditures on, effect of, 119–120; community control or influence of, 104–106; entrance-by-examination, 92–93;

ethnic groups free to establish own, 25, 26; private, 105

Scottish origin, Americans of, 173, 179

Seabury, Paul, 226*n*

Seattle (Washington), 170

Segregation: residential, *see* Residential segregation; of schools, *see* School segregation

Seligman, Daniel, 226*n*

Seniority rights, 35, 187

Shannon v. *HUD,* 158

Shelley v. *Kramer,* 133

Sherain, Howard, 211, 226*n,* 234*n*

Singer, Isaac Bashevis, 183

Skelly-Wright decision, 35

Slavery (slaves), 6, 10, 12, 207; justification of, 14–15

Slavic origin, Americans of, 191

Smith, Al, 21, 170

Smith, Marshall, 230*n*

SMSAs (Standard Metropolitan Statistical Areas), 132

Sobol, Richard, 226*n*

Social classes, *see* Classes, social

Social inequality, 30–31

Social relations, effect of residential redistribution of blacks on black-white, 147–148

Social service agencies of ethnic groups, 26

Solomon, Barbara Miller, 18, 19, 224*n*

Sørensen, Annemette, 231*n*

South, the, 3–4, 22, 40, 139, 212; consensus of the mid-1960s and, 5; exclusivism in, 21; opponents of racism do not want to adopt posture of, 209–210; Reconstruction in, 16, 17; school desegregation (busing for racial balance) in, 78–89, 93, 95, 98, 118, 206; voting rights for blacks in, 50; white backlash in, 169–171

Southern European immigrants and their descendants, 6, 18, 21, 75, 154, 155; census data on, 173, 175

Soviet Union, 22

Spanish Americans (Spanish-surnamed or Spanish - speaking Americans), 47, 103, 182, 184, 213, 228*n*, 229*n*; affirmative-action coverage of, confusion and inequalities concerning, 198–200; census data on, 173; *see also* Cubans; Mexican Americans; Nicaraguans; Puerto Ricans

Standard Metropolitan Statistical Areas (SMSAs), 132

Starrett City (New York), 140

States' Rights, 207

Stokes, Anson Phelps, 223*n*

Suburbs: affirmative action to achieve a more even racial distribution between central cities and, *see* Affirmative action in housing; black and white population trends in, 132–133, 136–143; effects of a more even distribution of blacks between central cities and, 146–151; houses built in, 132–133; jobs in, 131–132, 149, 150; policies for opening up the, 151–167; population and income data for, 132–133; residential segregation in, indices for, 143, 144

Sudman, Seymour, 231*n*

Supreme Court, U.S.: affirmative action and role of, 208, 216, 217; moral support for decisions of, 208; school desegregation and, 85, 87–90, 94–95, 113

Suttles, Gerald D., 189–190; 234*n*

Swahili, demands for teaching of, 184

Swann v. *Charlotte-Mecklenburg County Board of Education,* 87, 228*n*

Symbolic demands, 184

Taeuber, Alma, 142, 231*n*

Taeuber, Karl E., 142, 231*n*, 232*n*

Taggart, Robert, III, 228*n*

Teachers: quotas for hiring of, in Boston, 65–66; salaries of, black-white differentials, 120

Territorial classifications, as substitute for racial classifications, 117–118

Tests, ability, *see* Ability tests

Timetables and goals, for ending disproportionate employment of minorities, 36, 37, 46, 48, 58, 59, 211

Title VII of Civil Rights Act, 34, 47, 53; individual rights, not group rights, protected by, 44–45

Trade unions: employment discrimination and, 50, 178; later white ethnic groups and, 186–187

Truman, Harry S., 21, 46

Ukrainians, 27

Un-American Activities Committee, 216

"Underutilization" criterion, in affirmative action in employment, 47–49, 57, 66

Unemployment of blacks, 71, 72, 150–151

Unions: employment discrimination and, 50, 178; later white ethnic groups and, 186–187

Universal nation, Jefferson's view of America as a, 11–12

Index

Universities, *see* Colleges and universities

University of Massachusetts, 61

University of Michigan, Institute for Social Research of, 145

Urban Development Corporation, 161

Urban Economic Development (Harrison), 149

Urban Institute, The, 164, 225n

Useem, Elizabeth L., 230n

Vacation of the Kelwyns, The (Howells), 20

Validation of ability tests, 51, 52, 54–57

Voting Rights Act of 1965, 3, 22, 50

Wallace, George, 169, 170, 186

Washington, D.C., 140

Washington, George, 9

Washington State, 25

Wattenberg, Ben J., 225n, 233n

Weaver, Paul, 234n

Weigel, Stanley, 90, 91, 100, 110, 121

Weinstein, Jack B., 229n

Welfare, 72, 188

Welsh origin, Americans of, 173, 179

Wendell, Barrett, 224n

White ethnic backlash, *see* Political reaction of white ethnic groups

Williams, Harrison A., Jr., 45

Wilson, James Q., 230n

Wollheim, Richard, 207

Woodward, C. Vann, 223n

Wright, Skelly, 110

Yonkers (New York), 143

Youth, black, 71, 72

Zoning regulations, low-income housing and, 135, 148, 161–165